Mississippi Harvest

*Lumbering in the Longleaf Pine Belt,
1840–1915*

Nollie W. Hickman

University Press of Mississippi   Jackson

www.upress.state.ms.us

The University Press of Mississippi is a member
of the Association of American University Presses.

Print-on-Demand Edition
ISBN 978-1-60473-287-0
Published 2009 by University Press of Mississippi
Manufactured in the United States of America

Preface

Although lumbering played a role of major significance in the economic history of the South from the beginning of settlement to the present, it nevertheless failed to attract the attention of historians until recent years. University libraries and state archives only lately have begun to gather the documents from which a history of the southern forest products industries could be written. Most of the extant records of lumber or naval stores companies remained in the possession of business firms or heirs of the owners of enterprises no longer in operation and were therefore not readily available to scholars. Because of these conditions the historian either delayed writing the history of the southern lumber industry until the libraries and archives collected enough information to support a large-scale study, or else confined his approach to that of local history, making his own collection of documents.

Pursuing the course of local history, I have limited the scope of this book to the longleaf pine industries of the Mississippi Gulf Coast. This is an area small enough to permit a thorough search for records, yet large enough to serve as a representative sample of the whole southern longleaf pine industry. In order to keep the investigation within workable bounds, I have omitted other divisions of the southern lumber industry such as shortleaf pine, hardwood, and cypress, each of which is important enough to deserve a study of its own.

I should like to thank the libraries of the University of Texas, the University of Mississippi, the University of Tennessee, and Yale University for their assistance. I am indebted to many of the old-time loggers, rafters, sawmill workers, and naval stores workers for invaluable information. Walter Barber and Mayers Dantzler rendered particularly valuable assistance by permitting the use of voluminous papers of the L. N. Dantzler Lumber Company. Others too numerous to mention also contributed much to the project. I wish especially to acknowledge the help of Dr. Barnes Lathrop, professor of history of the University of Texas, and of Dr. John H. Moore, Dr. James W. Silver, and Dr. William H. Willis of the University of Mississippi, who offered many valuable suggestions and gave much of their time to editing the manuscript. I deeply appreciate the financial assistance extended by the University of Mississippi, the Mississippi Forestry Association, and the Forest History Society, which has made possible the publication of this book.

March, 1962   NOLLIE HICKMAN

Contents

Maps and Illustrations

MISSISSIPPI HARVEST

CHAPTER I

THE SETTING

THE LONGLEAF PINE BELT is one of the largest land divisions of the South Atlantic states. Its limits are set by soil types, moisture, and temperature. In width one hundred to one hundred twenty-five miles, bounded on the north by the foothills of the Appalachian Mountains and on the south by the coastal marshes, the longleaf pine belt extends from southeastern Virginia south and westward to the Trinity River in east Texas. The longleaf pine tree is not found in swamps or prairies or alluvial soils, and is limited mainly to orange sand or Lafayette strata or post-tertiary formations. In east Texas moisture conditions restrict the pine belt to an area south of the thirty-second parallel and east of the Trinity River. In Alabama, an area of heavy rainfall, the forests grow as far north as 34° 30' north latitude. The pine belt in Mississippi reaches from the Alabama border to the Bluff Hills in the western part of the state and extends approximately one hundred to one hundred fifty miles north of the Gulf of Mexico.

Within the pine country of Mississippi which lay between the Pearl and Pascagoula rivers, pure longleaf forest in 1880 constituted 75 per cent or more of the total land area. Other valuable pine trees found in the longleaf pine belt were shortleaf, loblolly, slash or Cuban, and bottom pine. Cypress, white oak, gum, hickory, poplar, magnolia, ash, beech, and other hardwoods of commercial importance, and cleared land accounted for the remainder of the piney woods. (See Map I.)

The general flatness of the land and the absence of thick undergrowth enabled the traveler to pass through the pine forests with as much ease as over a prairie. The pine region, where the burning of the woods occurred regularly, was almost devoid of shrubby undergrowth.[1] John F. H. Claiborne, writing in 1840, said:

> *For twenty miles at a stretch in places you may ride through these ancient woods and see them as they have stood for countless years, untouched by the hand of man and only scratched by the lightning or the flying tempest. This growth of giant pines is unbroken . . . for an hundred miles or more, save where rivers or large water courses intervene. . . .*[2]

Visitors were impressed by the miles upon miles of unbroken forests which, during the growing season, presented the appearance of a park. Many were depressed, however, by the unvarying uniformity of undulating hills and forests. Others born and reared in the piney woods were to experience a feeling of loss and longing for the forest world of their childhood.

Because of their unusual qualities, slash and longleaf pine, both southern trees, were two of the most valuable pines in North America. They were hard, possessed great tensile strength, and resisted decay. The natural beauty of the wood, combined with high resinous content, made them desirable for naval construction, framing, flooring, interior finishing, railway ties, spars, shingles, car sills, piling, telephone poles, and other purposes. In the antebellum period the versatility of the longleaf pine was thus described by Lewis Troost of Mobile, Alabama:

> But it is in the southern portions of the South and West where the most valuable timbers for naval architecture are found. Stretching through North and South Carolina, Georgia, Alabama, and Louisiana, are vast forests of the long yellow pine, of matchless height and straightness; it is one of the most useful and valuable of the forest trees, not only on account of the excellent quality of its timber, but for the tar, pitch and turpentine which it yields in great abundance.[3]

Longleaf is only one of twenty-seven or more names by which this pine has been known in world trade and in different localities of the United States. Some of its other names are Georgia pine, pitch pine, southern pine, yellow pine, North Carolina pine, heart pine, Texas yellow pine, and long straw pine. Longleaf is the name most generally used today because the length of its leaves best identifies the species. The length of the leaves varies from nine to fifteen inches, depending upon age of tree, fertility of soil, and moisture conditions.

The longleaf pine belt is the largest and commercially the most important forested region of Mississippi. Most of the longleaf country lies between the Pearl River and the Alabama border. From the Gulf Coast the longleaf pine extends northward through the counties of Copiah, Rankin, Smith, Lauderdale, Newton, Scott, and Kemper. Pure longleaf forests covering large areas originally were to be found on the northern border only in the counties of Copiah, Rankin, and Smith; and in the other counties the forests were confined to small sections frequently intermixed with shortleaf pine and hardwoods. Everywhere on the northern and western boundaries of the pine belt, the longleaf pine was gradually supplanted either by hardwoods or shortleaf pine or by a mixture of both. The central prairie, ten to thirty miles wide and running in a northwest-southeast direction from Warren County to the Alabama border, divided the longleaf country into two sections in Clarke and Wayne counties. The original forests on the crests of hills within the prairies were either longleaf, shortleaf, or hardwood.

West of the Pearl River, longleaf forests gave way to shortleaf and hardwood in northwestern Copiah, eastern Franklin, and about the middle of Amite counties. The southern and southwestern boundary of the forest west of the Pearl River was Lake Pontchartrain and the lowlands of the Amite River.

On the southern boundary of the pine country near the coast, slash pines grew in the wet meadows and in the lowlands along the streams. Much of the timber in the poorly drained, wet, sandy flats near the Gulf Coast was dwarf pine. By contrast, the quality and volume per acre of the timber located on the uplands a short distance inland often exceeded that in other areas.

The topography of the pine country is generally rolling, characterized by broad, dry plateaus occasionally cut by hollows and streams. On the east the rolling pine uplands of Alabama extend into Mississippi for several miles. West of the Pearl River the land in some areas is extremely hilly.

Throughout the entire pine country the surface of the soil is orange colored and sandy except in river and creek valleys. In many areas pure sand is found close to the surface. In other places hardpan and heavy soils are located a few inches below the top soils. Despite the apparent over-all uniformity of pine country soils, they are, nevertheless, unequal in fertility. In general the largest trees were to be found in the most fertile soils. Though occasional stands of timber west of the Pearl River ran to 30,000 board feet per acre, the average volume was between 10,000 and 12,000 board feet.[4]

In Simpson and Rankin counties and probably elsewhere, much of the timber was of exceptional size but also defective because it was too old. Such overripe timber contained a high percentage of redwood or deadwood, which made an inferior grade of lumber. In localities where occasional tornadoes or hurricanes had mowed down the forests, the timber was younger, smaller in size and grew in thicker stands. Where heavy clay soils prevented the trees from sinking their long taproots deeply into the earth, the timber was easily blown down.[5]

In the prerailroad era creeks, bayous, and rivers were of incalculable importance to the economy of the pine country. The Pascagoula River, which drains almost all of southeast Mississippi from Newton, Neshoba, Kemper, and Lauderdale counties to the Gulf of Mexico, was navigable for small boats. The Escatawpa River, originating in southwestern Alabama, forms a chain of natural lakes before joining the East Pascagoula River at Moss Point. These lakes, because of their depth and isolation from ocean currents, were ideal sites for forest industries. The Pearl River flows from its headwaters in Attala, Winston, Neshoba, Leake, and Kemper counties in a general southwesterly direction to the coast. Although a larger stream than the Pascagoula, and useful to the cotton growers on its lower reaches, the Pearl was of lesser importance to the pine country, since most of its important tributaries originated in shortleaf and hardwood areas. The Bogue Chitto River is one of the largest streams emptying into the Pearl River in the pine country. Other important streams, the Wolf, Jordan, Little and Big Biloxi rivers, empty into Mississippi Sound, and drain a large area between the Pearl and Pascagoula rivers.

From the Alabama border to the Louisiana-Mississippi boundary, the coast line was broken by numerous bays and inlets. Bay St. Louis, the Back Bay at Biloxi, and Pascagoula Bay offered protection for small vessels when tides and winds were high. Lake Borgne at the mouth of the Pearl River provided quiet waters and a relatively safe passageway to Lake Pontchartrain and New Orleans. Protection of the bays and inlets was bolstered by a chain of islands lying from two to twenty miles out at sea. Large ocean-going vessels could usually take on a full cargo in still waters behind island barriers. Occasionally, however, a southeast hurricane vented its full fury upon the coast country, bringing destruction to property and cutting a wide swath in the virgin forests of the interior.[6]

Farming in most of the pine country during the antebellum period was confined chiefly to first and second bottomlands adjacent to streams. Sandy ridges, plateaus, and hills lacked the natural fertility to support extensive cultivation. There were, however, a few upland areas where the forest growth consisted mainly of blackjack, post oak, and red oak and the soil was sufficiently fertile to be farmed. Only in these areas among the pine barrens, varying in extent from a few acres to several sections, were there upland settlements of any size.

Orange-colored, sandy, clayey soils consist of large numbers of subtype soils varying widely in respect to fertility, ability to retain moisture, and other characteristics. To produce crops, the most fertile pine country soils, such as those of the Orangeburg, Greenville, Norfolk, and Ruston soil groups, required application of foreign matter to some degree; and even today a majority of the subtype soils that constitute most of the land area of the pine country are not cultivated. In the ante-bellum period probably less than 2 per cent of the hill land was farmed.

The general unsuitableness of the pine country for profitable agriculture is indicated by the population statistics. The sixteen Mississippi pine counties with an aggregate of 103,269 inhabitants in 1860 were, except for the Yazoo-Mississippi Delta, the most thinly settled region in the state. Immigration had been fairly heavy prior to 1820, but in the decade 1830-1840 Greene, Wayne, and Perry counties declined in population while others grew but slowly. Farming at its best was risky and hazardous in the first bottoms, and on the hills large scale cultivation was impossible. The lack of fertility of the pine country soils motivated large numbers of farmers to move northward following the removal of Indians from middle and north Mississippi.[7]

On his leisurely travels through southeast Mississippi, Claiborne saw abandoned plantations and the once flourishing town of Winchester literally tumbling to pieces. Lawrence and Wayne counties had given the largest number of settlers to the new productive territory ceded by the Indians in 1830. B. L. C. Wailes observed in 1852 that there were only two houses fit for habitation in the once thriving town of Columbia.[8] Concerning Perry County in 1838, L. A. Besancon wrote:

> The population . . . is small in proportion to the square miles; when this part of the county was brought in to market it was settled with astonishing rapidity, and in a very short time became very populous; a few years, however, served to con-

vince its inhabitants of the uncertainty of their prospects;
and the extraordinary inundation of the lands, together with
the opening of a new purchase . . . from the Choctaw nation
of Indians, caused the tide of immigration to flow as rapidly
from the county as it had previously done towards it.[9]

Only in Copiah, Clarke, Jasper, Lawrence, and Pike counties, with their prairie soils and large areas of bottom lands, was population growth comparatively rapid during the period 1830-1860. Population increase was extremely slow in the pure piney woods counties. The increase for Perry from 1820 to 1860 was 569 people; for Greene, 787; and for Wayne, 368. In the three coastal counties population grew slowly but steadily, each succeeding census showing slight gains. This increase cannot be attributed, however, to the development of agriculture, for the sandy soils in the tidewater section were the poorest of the state. It was due primarily to the coming of industry and the development of the coast as a watering place for ante-bellum planters and New Orleans businessmen.[10]

Commercial agriculture was of small importance in the pine country in the first half of the nineteenth century. In 1849-1850 the whole area produced only 30,206 of about 485,000 bales of cotton grown in Mississippi; and of this small amount the three counties of Copiah, Lawrence, and Pike, having relative advantages in fertile soils, accounted for fully one-half. At the same time, Marshall County in north Mississippi produced more cotton than all the sixteen pine counties. In corn production the situation was very similar. Despite the production of some sea island cotton on the coast and the development of a few small plantations and farms along the lower reaches of the Pascagoula and Pearl rivers and in the bottom of Red and

Black creeks, in most of the pine country agriculture was of a subsistence nature.[11]

The people were stock-raisers and hunters. Their country was one great pasture where cattle and sheep fed upon wild oats and wire grass in the spring and summer seasons and upon reeds and cane of the bottoms during the winter months. Hardwood and pine mast provided food for large herds of swine. Throughout the woods were squirrels, turkey, and deer in great numbers. Grass and wild game were the two fundamentals which determined the basic economy of the pine woods folks for the greater part of the nineteenth century.[12]

The pattern of immigration to the piney woods was fairly definite. First came the hunters and herdsmen whose needs were simple and easily satisfied. These first settlers were described by William H. Sparks as having come from

> *the poorer districts of Georgia and the Carolinas. True to the instincts of the people from whom they were descended, they sought as nearly as possible just such a country as that from which they came, and were really refugees from a growing civilization consequent upon a denser population and its necessities. They were not agriculturists in a proper sense of the term; true, they cultivated in some degree the soil, but it was not the prime pursuit of these people, nor was the location sought for this purpose. They desired an open, poor, pine country, which forbade a numerous population.*
>
> *Here they reared immense herds of cattle, which subsisted exclusively upon the coarse grass and reeds which grew abundantly among the tall, long-leafed pine, and along the small creeks and branches numerous in this section. Through these almost interminable pine-forests the deer were abundant, and the canebrakes full of bears. They combined the pursuits of hunting and stock-minding and derived support and revenue almost exclusively from these. . . .*

Along the margins of the streams they found small strips of land of better quality than the pine-forests afforded. Here they grew sufficient corn for bread and a few of the coarser vegetables and in blissful ignorance enjoyed life after the manner they loved. The country gave character to the people: both were wild and poor. . . .[13]

In 1840 Claiborne learned while traveling through the piney woods of east Mississippi that many of the inhabitants owned large droves of cattle. Etienne Maxson likewise observed that

Cattle kept fat all the year around from the canebrakes before forest fires destroyed them. The bear, deer, and turkey roamed the forest; wild game was so plentiful up to the Civil War, that the deer had trails in the woods like cattle. All the hunter needed to do was to take a stand behind a tree near the trail and wait for his game. . . .[14]

Such conditions seem to have prevailed throughout the pine country of Mississippi. In George, Hancock, Jackson, and Harrison counties, local tradition supports the theory that hunting and herding were the chief pursuits of the pioneer economy. William Griffin, a resident of Perry and Jackson counties, purchased his first cattle with money received from the sale of dried venison in Mobile. His herds, ranging over a vast area from the head of Black Creek to the Pascagoula River, multiplied so rapidly that by 1850 Griffin counted his cattle by the thousand. According to Willard F. Bond, the pioneers who settled in the forks of Red and Black creeks in northwestern Jackson County near the Pascagoula River stopped in this area partly because of the plentiful supply of fish and wild game and because the cattle feeding upon the wild native cane stayed fat all year around. Wailes also commented upon the presence of large herds of cattle subsisting upon the grass and cane.[15]

The development of a grazing economy in the pine woods went through three well-defined phases. The migrating Georgians and Carolinians could bring only a few head of stock with them from their former homes. Before 1820 most of the journey lay through country where the danger from Indians lurked behind every bush and beyond each hill. These early settlers built their first homes on the creek bottoms. There, by unremitting toil, they cleared the hardwoods and made a few acres of bottom land available for subsistence crops.

The next step, a move from the swamps to the pine ridges, was made possible when the manure of the settlers' cattle had accumulated sufficiently to enrich a few acres for production of basic food crops. Three or four years of cultivation had by this time destroyed the accumulated humus of the sandy bottom lands. And by this time there was an awareness of the constant danger of crop destruction by overflows, as well as the belief that bottom lands were the cause of fevers.

This movement of settlers from the swamps to the hills can be followed in their first and second purchases of land. Entries by the early settlers in Harrison and Jackson counties were located near streams, but later acquisitions by the same people were on the uplands. The Griffin family, for example, settled first in the lowlands of Hickory Creek, later moved to the first bottoms on the south side of Black Creek, and then, after acquiring large numbers of cattle and sheep, moved a third time to the pine levels.

The migration of the Ward family from North Carolina to Mississippi likewise illustrates the practices of the early pioneers who settled first in lowlands and later moved to the ridges. The Wards settled near the junction of Cedar Creek and the Pascagoula River prior to 1810. There they cleared a small area of swampland and tended their live-stock. After a few years they moved with their herds of cattle to a more healthful location upon the nearby hills.[16]

The Ramsey family, who moved from Georgia to Mis-sissippi in the winter of 1807-1808, followed the same pat-tern. Its members settled first in the bottom lands of the Chickasawhay River in Wayne County. There, by clearing and selling bottom lands to squatters, they accumulated sufficient funds to purchase a considerable number of cattle and sheep. In the fall of 1819 the family drove their live-stock south of Red Creek and settled on a high bluff overlooking the stream. The region was thinly peopled; the closest neighbors lived four and eight miles away. The range was excellent in winter and summer for cattle, sheep, and hogs. The Ramseys cleared only a few acres of land, enough to provide vegetables for the household and feed for the horses.[17]

The third phase of settlement occurred when people moved from one locality of the longleaf pine belt to another. These migrations were sometimes dictated by the destruc-tion of the range through overgrazing and injudicious burning of the woods. Often men moved to another locality because the new country was more pleasing to the eye. Perhaps the most fundamental motive was the desire of the pine woods people to live in splendid isolation, far apart from their neighbors. A thickly populated area meant to them restriction of the range for livestock and destruc-tion of wild game, both striking at the heart of their means of livelihood. With the disappearance of grass and game, they turned to the tillage of the poor, sandy soil and to other occupations in order to satisfy their simple needs.

Writers of the period as well as oral tradition testify to the self-sufficiency of the pine woods people. According to family lore, A. W. Ramsey, who represented Harrison, Perry, Greene, and Jackson counties in the state senate in 1850, was adjudged by his colleagues to be the best dressed man in that body. Every stitch of his clothes except a silk shirt was made by his wife on a hand loom.[18] Claiborne spoke of the piney woods people as

> *living in a state of equality, where none are rich and none in want; where the soil is too thin to accumulate wealth, and yet sufficiently productive to reward industry; manufacturing all that they wear, producing all they consume; and preserving, with primitive simplicity of manners, the domestic virtues of their sires. . . .*[19]

In the late ante-bellum period there were many signs that the pastoral economy practiced by the piney woods people could not go on forever. Slowly but steadily population was increasing in almost every locality. The gradual disappearance of cane and reed was a catastrophe to cattle men, because these were the only food for the herds during the winter months when the tall grass under the pines was coarse and inedible. Deer and turkey, once abundant, were becoming scattered and shy before the hunter's rifle.[20] Eugene Hilgard, who traveled through many of the piney woods counties in 1860, stated that originally the pine country as received from the Indians was one great pasture. He noted that the pasture of the region, because of fires, was disappearing at a fearful rate:

> *those who heretofore have relied on the range, during all but a few weeks in winter, for the support of their cattle, will soon be compelled, as many are now, to raise feed for them on the poor sandy soil, which, at present, will but just furnish comfortably the prime necessities of life for the population*

itself. The beautiful park-like slopes of the Pine Hills are being converted into a smoking desert of pine trunks, on whose blackened soil the cattle seek more vainly every year the few scattered, sickly blades of grass whose roots the fire has not killed.[21]

The appearance of saws and steam engines on the banks of the rivers and bayous near the Mississippi Sound heralded the approach of a new economic order. This new system would completely alter the economy of grass and game. It meant that volunteer refugees from civilization would be literally thrown into an environment from which they had tried to escape by retreat. The desire of the piney woods people to live out their lives far removed from a complex society had motivated their often dangerous travels to the lonely pine woods. But isolation and a simple existence dependent upon the prevalence of grass, game, and a small population were coming to an end. The people henceforth would become irrevocably bound to the industrial age, most of them as underpaid wage hands. Their land of beauty, the fulfillment of all their simple requirements, would fade into a nostalgic remembrance, with no one to understand the full meaning of the words "pine barrens."

CHAPTER II

ANTE-BELLUM LUMBERING

For countless generations the vast forests of the long-leaf pine belt of Mississippi remained almost unbroken. The pioneer herdsmen and hunters removed the timber from only a few acres. To them the trees as such had no value except as wood for their crude houses, furniture, fences, and plows. The pines with their long taproots represented major obstacles to clearing the few acres of farmland upon which the settlers were dependent for subsistence crops.

From the earliest years of the white man's occupancy of the tidewater, timber in small amounts had been taken from the forests. The early French colonists had felled the tall, straight, tapering longleaf and slash pines and converted them into masts and spars for sailing vessels. The practical French colonial officials attempted to develop forest industries in the barren coast country, not only to make the colony self-supporting, but also to furnish the Sugar Islands and other French possessions a much needed commodity. All along the piney woods coast, small sawmills were erected by the French and later by the English and Spanish colonists. The Spaniards carried on both lumbering and the naval stores business. William Bartram noted in his trip through West Florida the existence of huge iron pots used in the distillation of tar, pitch, and turpentine.[1]

In the operation of an early sawmill, timber was roughly squared with a broad ax and then placed over a saw pit or elevated on a trestle six or seven feet high. Saws of several patterns were used; the most common had a blade six or seven feet in length. A sawyer stood on the upper side of the log and pulled the saw upward, while the pitman stood underneath and pulled the saw downward. The saw cut only on the down stroke. Another important type, the sash sawmill, appeared in the late colonial period and continued in use until the 1840's or later. The saw consisted of a vertical blade fixed in a rectangular frame which gave stability to the sawing operation; the weight of the frame limited the number of strokes made by the saw to 120 or less per minute. A crew might saw from 3,000 to 5,000 feet daily. Both saws made a kerf of more than one-half inch, but such waste was of no concern at a time when the supply of timber was deemed to be inexhaustible.[2]

For the first two American decades wood industries differed little from what they had been in the colonial period. A few mills were erected in the coast country where water power was available and in the back country where an occasional settlement created demand for a small mill. The interior sawmills were commonly operated in conjunction with grist and rice mills and cotton gins. Commerical lumbering awaited the coming of the steam engine and the development of markets.

Just when and where the first steam sawmills were erected along the coastal waterways is a matter of conjecture. During the middle thirties, however, both markets and technology invited the development of commercial lumbering. Already the superior qualities of the longleaf and slash pines were known in such far-off places as England and Western Europe.[3] Within a few years agents of the French government made their way to the Gulf Coast in search of spars and square timber for naval vessels. Cuba, Mexico, and Texas, in close proximity to the Mississippi Coast, also provided early markets for lumber. But the earliest market for lumber arose at nearby New Orleans. Here population, trade, and industry grew rapidly in the years 1830-1850. The coast country of Mississippi, lying within a short distance of the city, could supply all the forest products needed for an expanding market.

The first steam mill on the Mississippi Coast was said to have been erected at Pascagoula in 1835 but was apparently operated for only a short time. By 1840 there were ten sawmills in operation in Hancock County and in that part of Hancock which became Harrison County in 1843. Jackson, the other coast county, had only two small mills in 1840, while at the same time the backwoods county of Lawrence boasted ten.[4]

Because logs had to be brought to the mills by water from interior forests, and lumber shipped to outside markets by boat, almost all of the early mills in the coast country were erected at river mouths or on the banks of bayous which extended a few miles into the interior. Indeed, before the advent of the railroad, commercial lumbering would have been impossible without cheap water transportation. In Hancock County the mills were a short distance up the Pearl River from Lake Borgne and at the head of the Bay of St. Louis. Pearlington, Napoleon, Logtown, and Gainesville, located on the Pearl, were early sawmill sites. Far into the back country the logs were cut and hauled to the Pearl and its tributaries, and then floated downstream to the Pearl River mills. There the logs, both cypress and pine, were manufactured into lumber, staves, and shingles and shipped by schooners and brigs to the markets outside. The Pearl River mills possessed an unusual advantage in that New Orleans was less than a day's journey away by water.[5]

One of the earliest of the Pearl River lumbermen was W. J. Poitevent, who came to Gainesville in 1832 from the Orangeburg district in South Carolina. Poitevent probably erected a mill shortly after his arrival. By the middle forties he had become a wealthy man according to the standards of his time. He conducted a number of enterprises that included sawmilling, shipping, and storekeeping. His business expanded with the years, especially in sawmilling and the construction of schooners. In 1860 Poitevent was the owner of two sawmills, one at Pearlington and the other at Gainesville, and he subsequently acquired large timber holdings in both east Louisiana and Mississippi. Following the Civil War he owned for a number of years, in partnership with Joseph Favre, the largest sawmill in the coast country.

As associate of Poitevent was D. R. Wingate, who migrated to Pearl River in 1824 from the Darlington district of South Carolina. The details of Wingate's early ventures are unknown; but in 1844 he was operating a sawmill in partnership with Poitevent, and he continued active in the lumber business to 1856. The first jointly-owned mill was destroyed by fire, but another one was erected at Logtown. This partnership was dissolved by mutual consent in the late forties. Subsequently in 1854 Wingate formed a partnership which included W. W. Carré and his brother and Henry Weston. In 1856, when the Carrés and Weston purchased Wingate's interest in the mills, he moved to Texas where he built a large and prosperous mill business at Sabine Pass.

In the late forties and early fifties, a time when many lumbermen were migrating from New England to the virgin white pine forests of the Lake states, a few came to the Gulf Coast. Henry Weston was one of the newcomers. Born in 1823 at Skowhegan, Maine, he had become familiar, as an employee of his father, with every phase of the lumbering business. Attracted by stories of the great forests in the Lake states, Weston at the age of eighteen moved to Wisconsin, where he ran logs on the Eau Claire River for a few seasons until the rigors of the climate broke his health. To regain physical strength, he came south, arriving at New Orleans in 1846. In a strange country, penniless and without experience in any trade except the lumber business, he went to Gainesville to seek employment in the Pearl River mills. For the next few years Weston worked as a sawyer at $45 a month in mills owned by Poitevent and Wingate. Having demonstrated his ability to manage sawmill operations, Weston was given charge of two sash mills that had a daily capacity of 3,000 board feet.

In 1856 Weston and the Carrés purchased on credit a one-third interest in the Wingate mills valued at $15,000. Shortly afterwards the mills went up in flames, but the partners immediately began the task of rebuilding. Their new sash mills with a daily capacity of 9,000 board feet were finished by good fortune at a time when lumber prices were increasing sharply.

The partners set aside $5,000 annually for living expenses.

Remaining profits were turned back into the business. They bought the largest and best quality pine and cypress logs for $3 per thousand board feet log scale. In their mill almost all the laborers except the foreman and sawyers were Negro slaves. Profits were large in proportion to capital investment and labor costs. Production and marketing costs were much less than $9 per thousand board feet, and the selling price of lumber was between $10 and $20 per thousand on the New Orleans market. Referring to his lumber business in the late fifties, Weston stated that he made money like smoke. This prosperity enabled Weston and the Carrés to lay the foundation for a large business immediately after the war.[6]

There were other mills in the Pearl River district. From May 31, 1849, to June 1, 1850, mills in Hancock County sawed 30,000 to 40,000 logs into approximately 7,000,000 board feet of lumber. Figures for lumber production do not include square timber brought down the streams in unknown amounts to the coast for overseas shipment. D. R. Walker and Toulme owned a mill valued at $9,000 that sawed 1,200,000 board feet. Wingate's mill employed fourteen workers; its production was valued at $15,000. W. M. Brown, a Shieldsborough lumberman who employed eleven men and two women, grossed $11,200. A. S. Hursey of Maine and his partner Henderson operated another small steam mill that sawed about 700,000 board feet of lumber.

While little information is available on lumbering in Hancock County in the years 1850-1860, there is some evidence that sawmilling continued to expand. B. L. C. Wailes in his trip through the county in 1852 reported eight mills at Pearlington and two other large ones in the process of erection. Professor Trowbridge of the United States Coast Survey noted around the shores of the Bay of St.

Louis seven mills from which 1,000,000 board feet of lumber were shipped annually. The French government purchased in the Bay area both lumber and large supplies of spar timber for naval construction. Lumber and timber from all along the coast were shipped to many parts of the world including distant Australia.[7] By 1860 the economy of the inhabitants on the lower reaches of the Pearl and Jordan rivers rested almost entirely on the products of the forest.

Lumbering at the month of the Pascagoula River developed slowly before 1850 in spite of the unusual advantages of the site, lagging far behind the Pearl River and Bayou Bernard districts. One of the first mills was located at Americus some distance up the Pascagoula River from the coast. This mill, built in the early thirties and owned by Sark and Dameron, was a combination sawmill and gristmill. Its small production was marketed down the river to the residents of the Pascagoula area. One of the early steam mills in the district was built by Tetar at West Pascagoula in 1835. Its existence was brief, for Tetar failed, the mill was dismantled, and its engine installed in a boat. Perhaps the first mill in the neighborhood which became Moss Point (known earlier as Mossy Pen Point, a shipping port for cattle) was built by Beardslee and John Bradford in 1836. The site of the Beardslee and Bradford mill was on the bank of a natural lake of the Escatawpa River a few hundred yards above the confluence of the Escatawpa and East Pascagoula rivers. It was perhaps the best mill site in all the coast country; logs could reach the mill from both the Pascagoula and Escatawpa rivers, and schooners from the Gulf could take on cargoes at the mill. Beardslee and Bradford sold their mill to David Files, a native of Maine, who after operating it for a short time sold out to Walter Denny and Son, native Mississippians, in

1853. Files remained in the lumber business and erected another mill nearby.

On the Pascagoula, as on the Pearl, New Englanders were among the pioneer lumbermen of the district. Three Massachusetts men, J. M. and J. P. Arnold and W. M. Sheldon, in 1847 commenced the manufacture of lumber near the location of Beardslee's earlier mill.[8] The partners operated a modification of the sash mill known as the muley mill in which the crossheads held the saw blades in position, thus eliminating the heavy wooden frame that had been used to give stability to the blades of the sash saws.

In 1849 a fourth partner, William Griffin, joined the firm heretofore composed of the two Arnolds and Sheldon. Out of Griffin's interest in the mill business grew one of the largest lumber concerns in the South. His previous experience in lumbering had been confined largely to rafting spar timber and logs down Black Creek and the Pascagoula River to tidewater. From this business and his large herds of cattle, Griffin had acquired extensive holdings of land and slaves and perhaps surplus cash. His experience in logging and rafting was undoubtedly a valuable asset to the firm, set up as a ten-year partnership.[9]

Griffin left his farm in Perry County and came to Moss Point in 1850 or 1852. It appears that after the death of one of the Arnolds in the explosion of a steam boiler, Griffin took over the management of the mills, while his partners operated the marketing end of the business. In 1858 the firm increased its capital by the purchase of a gang saw-mill, constructed by adding saws to the muley saw so that more than one board could be sawn from a log in a single operation. This gang mill, with modifications, continued to be used until 1905.

The interests of the firm were far-flung. In addition to land and other property, it owned retail lumber yards in

Boston and a line of schooners used to transport its products to market. As was then common, the partners operated a brick kiln in connection with the sawmill, using slabs and refuse to provide fuel.

The partnership was dissolved in 1860. The northerners, aware that the Civil War was at hand, disposed of all their property to Griffin for $60,000. In winding up the business Griffin assumed responsibility for all debts and after all business had been settled, profited to the extent of $10,309.75. Thus in 1860 Griffin was in an excellent position to expand operations, since he owned mills, ships, timber, and labor.[10]

Malcolm McRae purchased a small mill that had been erected in 1845 by D. A. Vermillion at the mouth of the East Pascagoula River and operated it until 1860. On Red Creek, a tributary of the Pascagoula River, McRae owned a farm and a number of slaves. He employed them for most of the year to cut, haul, and raft timber to his mills at East Pascagoula. Even before entering the mill business, McRae had cut and rafted spar and square timber down to tidewater.[11] In 1850 his mill sawed 700,000 board feet.

The largest mill owner in the Moss Point-Pascagoula district in 1850 was J. S. Dees. His mill, the only watermill in the district, was located a few miles up the Escatawpa River on Jackson Creek and employed twenty-five hands including five women. Employment of women sawmill workers was not unusual in the ante-bellum period. From May 31, 1849, to June 1, 1850, Dees's mill sawed 2,000,000 board feet of lumber. William Deggs, a millman of Moss Point, cut 500,000 board feet of lumber in the same year. Total lumber production for the Pascagoula district in that year cannot be accurately estimated, for the census returns did not include the mill owned by Griffin and his partners and others mentioned by Mark Dees. According to the return, from May 31, 1849, to June 1, 1850, six small Jackson County mills sawed 19,700 logs into 5,650,000 board feet of lumber.

In the years following 1850 the number and capacity of the individual mills increased sharply in the Pascagoula district. A major reason was the adoption of the circular saw, first by Thomas and Rufus Rhodes and soon after by Garland Goode and John S. Dees.[12] The circular saw was the first continuously cutting saw, and it was the most significant development in lumber manufacture since the introduction of the sash mill. Although this saw was in use before 1850, not until the invention of the inserted tooth in the late 1840's was it made practical for sawmilling. Even then the circular saw had not reached its fullest development, for no satisfactory method of holding the teeth in place had yet been devised. In the 1860's curved sockets solved the problem. The advantage of the circular saw lay in speed, for it made from 7,000 to 10,000 revolutions per minute, and its use greatly increased the capacity of the individual mill.[13]

Garland Goode's mills, located above Franklin Creek on the Escatawpa River in the late fifties, were the largest and best managed in the district. Employing both circular saws and a double gang, the mills averaged about 25,000 board feet per day. The double gang was basically a combination of sash saws in a frame. A planing mill was added to the Goode sawmills; according to Mark Dees it was the first to be built in the Pascagoula district. Goode's mill management resembled the division, organization, and specialization of labor that was to prevail in large mills toward the end of the nineteenth century. He divided the mill into departments and placed a manager over each.

Most of the common laborers were the slaves owned by Goode; his only hired employees were department managers. Every phase of lumbering from the forest to the market was conducted by the millman. The cost of timber delivered to the mills was $2.48 per thousand, and sawing and shipment of lumber to market brought total costs to $5.00. With the high prices that prevailed in the late fifties the Goode business appears to have been prosperous.

In 1853 Walter Denny erected a circular sawmill on the Escatawpa River and began a mill business that continued beyond the end of the century. Like Goode, he owned his labor and transported lumber to New Orleans in his own vessels. Other millmen in the Pascagoula district were the partners Plummer and Williams.[14]

Prior to the Civil War, lumbering in the back country remained relatively undeveloped. In the vast areas where rail transportation was nonexistent, lumbering in 1860 continued as it had been in 1840. During the ante-bellum period small back country mills, far removed from lines of transportation, were erected in localities where population was somewhat heavy. These usually were combination mills, powered by water, which ground corn, ginned cotton, and sawed lumber in small quantities. There were fifteen water mills in Pike County in 1852. Eleven of them had cotton gins attached and the same number sawed lumber. All the mills ground corn meal and five were equipped to clean rice. Of the nine water mills in Covington County, five had saws and seven were combination cotton gins. Nearly all of the Covington County mills cleaned rice. In 1850 there were six combination mills, an indication that both population and wealth were increasing in this county. The average annual capacity of each mill was less than 500,000 board feet. In Copiah County, likewise, water mills sprang up in populated localities. The only commercial mill of any importance in the back country was one in Rankin County that turned out products valued at $9,000 in 1850.[15]

With the building of the Mobile and Ohio and the New Orleans, Jackson, and Great Northern railroads, commercial lumbering on a small scale arose in the interior of the longleaf pine belt. The two roads, one in the eastern section of the belt, the other on its western borders, penetrated the pine country in the years 1852-1857. Steam mills were erected along the paths of the roads, and their products were transported by the railroads to regions beyond the longleaf pine belt. In 1860 the Sanders brothers of Clarke County cut 2,000,000 board feet. Another steam mill owned by Harmon and Company also manufactured 2,000,000 feet. Moffett and Dyers of Clarke County cut square timbers in their steam mill valued at $13,000. This square timber probably went to Mobile, there to be sent overseas. On the western section of the longleaf pine belt, commercial lumbering in Copiah and Pike counties developed with the coming of the railroad. In Copiah, where only water mills operated in 1850, eleven steam mills had been erected by 1860. To the south in Pike County, W. W. Vaught erected a circular saw steam mill in 1855, perhaps the first of its kind in the county. J. J. White and his brother Robert Emmett White in 1859 started a lumber business that was to continue for more than sixty years in Pike County.[16]

The coming of forest industries to the mouth of the Pearl and Pascagoula rivers and to the interior, small as they were, signified the beginning of a shift from an economy based upon game and grazing to lumbering. All along the coast, especially at the mouth of streams, population and wealth increased with the establishment of forest industries. Although lumbering and other forest industries were in their infancy in the ante-bellum period, these early ventures foreshadowed the development of a large industry that would transform the economy and social order of an entire region.

CHAPTER III

ANTE-BELLUM LUMBERING
IN HARRISON COUNTY

THE LUMBER INDUSTRY at Handsboro located on Bayou Bernard and at Delisle reached a high peak of development in the late ante-bellum period. The Bayou Bernard area was known in the 1850's as the principal center of lumbering within a hundred mile radius of New Orleans. In volume of production, number of mills, capital invested, and related industries, the district was further advanced than other sections of Mississippi prior to 1860. Millmen had been attracted to the area early, mainly because long-leaf and slash pines grew almost up to the banks of its streams. In many places logs could be rolled into the stream by hand, so that logging was relatively inexpensive.

Almost all of the mills in the Bayou Bernard area were located within eyesight of one another on either bank of the bayou. At Buena Vista (later known as Handsboro) where fingerlike projections of land extended outward into the bayou, small eddies of deep water were formed. These areas, protected from incoming tides, served both as natural storage ponds for logs and as anchorages for lumber schooners. Beginning in the late thirties, lumbering developed slowly at first, then rapidly in the years 1845-1850. John J. McCaughan, a planter from Kentucky who lived at Palmetto City a few miles west of Handsboro, reported a considerable commerce in pine plank, charcoal, and cordwood. The cordwood went to the mouth of the Mississippi River to be used as fuel for boats employed in towing steamers on the river between New Orleans and the Gulf. Fernando Gautier, a French immigrant, operated a sawmill on Tchoutacabouffa Bayou in 1844. In 1846 Calvin Taylor, a native of New Hampshire, began a lumber business in conjunction with others that continued for almost forty years. In the latter part of 1846 and early 1847 a number of sawmill men and manufacturers of sawmill equipment arrived in the Bayou Bernard area. Among the immigrants were two Swiss, Henry Leinhard and his relative, Jacob Salmen, who would in time accumulate large fortunes from lumbering in this unfamiliar subtropical coast country. In the years 1846-1848 many New Englanders came to the Mississippi Gulf Coast. Indeed, so many of the skilled and professional classes in the Bayou Bernard district were of New England origin that Mississippi City came to be known in the late ante-bellum period as a New England community.[1]

Immediately east of what is now the Handsboro Bridge on Bayou Bernard, separated from it by a small peninsula, were two mills owned by Samuel Fowler of New York and Calvin Taylor. The two mills sawed 12,000 logs in the census year 1850. Across the bayou the firm of Hand Brothers and Prother sawed 6,000 logs. In connection with their sawmill the brothers operated a foundry. The Hand brothers had come from New York to the coast in late 1846 or early 1847. H. Bingham of New Hampshire, whose mill was on the south bank of the bayou, sawed 8,000 logs in 1850. Nimwood McQuire, born in South Carolina, operated both a steam and a water mill, producing 2,100,000 board feet. At nearby Cedar Lake, S. S. Henry, born in New York, sawed 12,000 logs.[2]

On the opposite or eastern shore of the Bay of St. Louis, at Delisle in Harrison County, where the Wolf River empties into the bay, another small lumber center grew up. Thomas Gray, G. L. Thomas, Leonard Staples, John Huddleston, W. J. White, and other millmen operated there or elsewhere in Harrison County. The combined log consumption of the seventeen Harrison County mills, which included both the eastern shore of the Bay area and the Bayou Bernard district, was 87,900 logs in the twelve-month period covered by the United States Seventh Census of 1850. The total production of lumber of all the mills, approximately 17,000,000 board feet, was extremely low in proportion to the number of logs sawed. In this period, when all salable sawmill timber had to be eighteen inches at the smallest end, clear of knots, and all heart, it is probable that more than one-third of each log was converted into kerf and slabs.[3]

Partly because the ante-bellum mills utilized only superior timber, the specter of a shortage of raw materials for the Bayou Bernard mills had appeared by 1850. The number of commercial trees per acre, according to methods of utilization then current in the coast country, averaged between five and fifteen. Thus the twelve-months' cut listed by the census of 1850 had removed the salable timber from more than 6,000 acres. Moreover, the streams upon which the mills were dependent for the transportation of their timber extended only a few miles into the interior. As long as the most efficient method of hauling logs to water was by caralog and oxen, it was uneconomical to bring timber more than three miles to the rafting stream. Obviously, the Bayou Bernard lumbermen, if they were to survive, had to develop another method of bringing their raw materials to the mills. As a solution to the problem, the millmen projected a railroad northward from the Biloxi River through a virgin forest to Red Creek, where the road would tap a vast area bordering on the creek and its tributaries. The millmen Calvin Taylor, Samuel J. Fowler, Donald McBean, Miles B. Hand, Henry Williams, Micajah Felton, James Brown, and others in 1852 secured a charter of incorporation to construct either a single or double track railroad.[4]

In that year the legislature of Mississippi memorialized Congress to shift the custom house from Shieldsborough (now Bay St. Louis) to Biloxi. There were, according to the memorial, one hundred steam and sailing vessels at Biloxi. Much of the timber was shipped to Cuba and Texas. It would be of great importance to both shippers and owners of vessels to have the custom house located at Biloxi, near the mills, to obtain clearance for foreign ports without delay.[5]

As a matter of fact the Gulf Coast mills at this time exported abroad only a small part of their total production. Most of their lumber went on consignment to wholesalers or commission men in New Orleans. In some instances it was shipped to New Orleans merchants to be sold for what it would bring on the market.[6] This system of marketing continued in effect almost to the end of the nineteenth century.

During the 1850's strong competition and unsatisfactory marketing arrangements produced what may be termed a lumbermen's association among the millmen of the Bayou Bernard. This was a prototype of the many organizations created in the late nineteenth and early twentieth centuries to control prices and eliminate the middle man wherever possible. Henry Williams from Maine appears to have been the guiding hand among the millmen. He had previously studied the market carefully and had obtained premium prices for his lumber in the New Orleans market. In 1852 he and other lumbermen of the Bayou Bernard area formed the Bayou Bernard Lumber Company. For a short period the company was successful in fixing prices, selling lumber directly to the consumer, and eliminating competition. But after a few months it fell apart, ending the first ambitious effort to establish control of the retail market through a producers' organization.[7] It is probable that competition on the New Orleans market from other lumber centers such as Pensacola and Mobile prevented the Bayou Bernard Company from achieving a monopoly of the lumber trade. Meanwhile lumber prices were rising and foreign trade increasing, developments which no doubt alleviated the dependence of the millmen on the New Orleans market.

In the years 1850 to 1860 lumbering continued to grow. In 1852 B. L. C. Wailes, after visiting Handsboro, reported an extensive settlement and many people working in the sawmills. Also a short distance to the east on the north shore of Back Bay were situated a number of sawmills. In the same location was perhaps the largest brick kiln in the coast country, owned by George Kendall from Kentucky. His steam-powered kiln represented an investment of $42,250 and produced annually 10,000,000 bricks valued at $60,000. Kendall, the largest slave owner in Harrison County in 1850, employed 116 men and 37 women in his plant.[8]

Lumbering on the Mississippi Coast was stimulated by growth of foreign and domestic demand. Although lumber, square timber, and spars had been shipped directly from mills on the seaboard to Europe and the West Indies probably as early as 1840, by 1856-1857 the volume of shipments became impressive. J. F. H. Claiborne in a letter to the editor of the *Mississippian* in 1857 stated that the mills located on the margins of the small streams which debouched into the Gulf annually supplied forest products to fifty cargo vessels from all parts of the world. During the latter half of April, 1857, lumber, sawed timber, and deals to the value of $28,000 were shipped to England and Australia. "In addition," wrote Claiborne, "our trade in lumber coastwise, that is to say with Texas, Mexico, and the West Indies is enormous," and "Mississippi pine is now sent on almost every steamer that leaves New Orleans for St. Louis."[9]

For a long time England obtained her best quality of pitch pine from Savannah. According to a writer in 1858 an excellent quality of pitch pine was now obtained from Ship Island. Each log was square sawn, and sent over in scantlings averaging twelve inches, and in lengths of twenty to thirty feet. In 1859 another writer stated that the lumber trade employed a large capital of a million dollars or more and that many people in it were accumulating great wealth. The transportation of lumber, cordwood, lightwood, tar, and shells employed more than a hundred sloops, barges, and schooners.[10]

Immediately before the Civil War the capacity of the individual mills in the vicinity of Bayou Bernard was increased by the introduction of the latest technological improvements in lumber manufacturing, such as circular saws and planing mills. Many newcomers migrated to bayou country and erected mills, among them J. T. Liddle, Marvin Holcombe, James Fewell, L. T. Burr, W. J. Blackman, and L. M. Taylor. In the bayou district there were a number of sawmills in close proximity, and in the Bay of St. Louis or Delisle area as well.

In a sense, Bayou Bernard was a center for the lumber industry from Florida to Texas. It was here that steam engines and boilers used along the Gulf Coast were manufactured. The L. N. Bradford foundry, consuming annually 175 tons of pig iron, manufactured $80,000 worth of engines and boilers. In the same year S. B. Hand, a competitor of Bradford, produced twenty-five engines valued at $62,500. The Bradford and Hand foundries together employed seventy workers.[11]

The development of the lumbering business is illustrated by the ventures of Calvin Taylor.[12] Born at Hindsdale, New Hampshire, Taylor came to Yazoo County, Mississippi, in 1826. There for twenty years he speculated in land and operated a plantation and a sawmill. In 1845 or early 1846 he formed a partnership with Gordon Davis, another New Englander, to purchase a small steam sawmill on Bayou Bernard. During the first few months the business seemed definitely unsuccessful. The foreman of the mill proved to be both ignorant of machinery and unable to manage the slave labor employed in the mill. The partners also lacked equipment and means to supply their mill with logs; apparently, they depended at this time on logmen for their timber supplies.

Despite the bad start, Davis was optimistic and willing to increase the size of his investment in the business. He travelled northward in the summer of 1846 ostensibly to improve his health but actually to inform himself of the latest improvements in sawmill machinery and to obtain skilled mechanics as operatives. Already dissatisfied with the small capacity of the Mississippi mill, he recommended the purchase of a newly patented planing mill which would cost the partners $300. He described another machine priced at $50 that was designed to saw either shingles or laths. A steam boiler to generate power for the machine might be had at Youngstown, New York, for $900. A new type of mill known as the belt mill interested Davis. It was not a common frame mill, for it could be attached to any power that had a pulley. If a planing mill were purchased, as he strongly recommended, it would become necessary to add a second mill in order to keep the planing mill busy. Davis's letters breathe a spirit of optimism regarding the future of the lumbering business on the coast. "The piney woods coast," said Davis, "has an attraction for me that I cannot account for." He was willing to expend additional capital for machinery and even more eager to secure trained mechanics and skilled sawyers. He thought the small output of the mill might be trebled if a skilled sawyer could be hired. At Oswego, New York, he met three sawyers who volunteered to come to Mississippi if they could be guaranteed $2 per thousand feet for sawing lumber. Whether Davis succeeded in obtaining skilled labor for his mill is not recorded.

Davis and Taylor continued to suffer losses in the lumber business. Between August and December, 1845, they sent to New Orleans 128,978 feet of lumber and sold it for $885.75, which was $112.17 less than cost. The grades of lumber being manufactured at the mill were scantling, ceiling, plank, rough edge, weatherboarding, and joists. The partners also engaged to some extent in selling on the New Orleans market lumber which they purchased from other bayou millmen. Like most lumbermen at that time, they conducted a general merchandising business in connection with the mill.

The mill business remained largely unprofitable for Taylor and Davis in 1847. In January of that year, for example, 22,200 feet were shipped to New Orleans, but less than one-half of the shipment could be sold. The partners drew a cash advance of $12 per thousand against the unsold lumber. While this price appeared adequate, a further loss of $500 was suffered by the firm within the next two months. The New Orleans price at that time was lower than $9 per thousand feet. Their losses continued, for in April and May they were $400 further behind. It is possible that Taylor recouped at least part of his losses in the mill business through sale of general merchandise to the logmen from whom he purchased timber.

In the latter part of 1847 and early 1848 Taylor's mills were running smoothly. When Davis died in 1848, Samuel Fowler became a part owner of the mills. Lumber prices averaged from $10 to $12 per thousand feet. Taylor and Fowler enlarged the capacity of their mills by the purchase of a circular sawmill from Hand. Later in 1849 and in 1850 lumber sales were slow. Prices declined: Charles Ludlow, a New Orleans commission merchant, informed Taylor that the depressed conditions of the lumber market were due to overproduction. In a second letter Ludlow informed Taylor that prices were continuing to fall. Because of the forced sales of lumber, Taylor had in June, 1850, 100,000 feet of lumber that found no sale.

Taylor disposed of his mill business about 1852 and moved to New Orleans, where he served for a short period as one of the agents of the Bayou Bernard Lumber Company. With the failure of the company, Taylor returned to Mississippi City to re-enter the mill business in partnership with Gordon Myers. This firm was to continue until the Civil War without change. One of the first moves after the formation of the partnership was the purchase of a planing mill, probably the first of its kind in the Bayou Bernard area. The machine, bought from Elisha Bloomer of New York, cost $500 payable in installments. The partners in their contract with Bloomer agreed not to manufacture the machine. With the acquisition of a planing machine Taylor and Myers secured an important advantage over their competitors. Dressed lumber not only commanded a higher price than rough lumber but also weighed less and could therefore be shipped to market more cheaply.

During the next few years the record of Taylor and Myers is obscure. When in 1856 their productive capacity was increased by the purchase from Hand of a second circular mill, the firm entered a period of unprecedented prosperity lasting until 1861. By 1858 lumber, especially dressed lumber, sold for prices ranging from $16 to $30 per thousand, while the cost of production remained about the same as in the decade 1846-1856. During the years 1856-1860 Taylor began to acquire timber land and more slave labor. From another sawmill owner Taylor purchased 1,350 acres. Tracts ranging from 40 to 160 acres of swampland were bought from Harrison County for five cents an acre. It would appear that premier timberlands were becoming scarce on the nearby streams, and millmen were beginning to buy lands to insure themselves of a future timber supply. At a public sale Taylor purchased part of the property of his former partner, Samuel Fowler, consisting of five Negro slaves, a saw, and a planing mill. The ownership of the Fowler mill made the lumber firm of Taylor and Myers the largest producer of lumber in the Harrison County area. In the census year 1860 the firm operated two sawmills which turned out 3,000,000 board feet of lumber and a planing mill that manufactured 1,700,000 board feet. The firm employed twenty-five workers in mill operations and had a gross income of $48,000.

The mills owned by Taylor were located, like all other mills of that period, on the margins of streams and bayous where the timber could be rafted easily and from which lumber could be shipped on schooners and vessels to the markets outside. Taylor procured most of his mill timber from logmen who cut timber, hauled it to the streams, and rafted it to the mills. Many of these logmen employed a considerable labor force in their business. In most mill operations the logmen contracted with the millman for the delivery of logs at a stipulated price. The millman, to insure a supply of logs, often made advances in cash and merchandise to the logmen. It was for this purpose, and to serve the labor employed in the mill, that almost all saw-mill owners operated a commissary.

Taylor and Myers, after having acquired considerable tracts of timberland located in close proximity to the rafting streams, began to supply their mills with logs cut from their own land. Nevertheless, a contemporaneous increase in the capacity of their mills left them still dependent upon the logmen for the greater portion of their logs. The dealings between Taylor and Myers and Goodman Hester, a logman who lived upon Tuxechaena Creek, reflect the relations between mill owners and logmen. Hester sold logs to the firm continuously from 1858 to 1861. In August, 1858, he borrowed $300 from Taylor to enter the log business. As a condition of the loan Hester was required to sell logs only to his creditor. Besides the loan Hester received advances of corn, coffee, flour, soap, tobacco, lard, oil, and other items charged against future log delivery. He bought corn of two different classifications: corn for Negroes and corn for livestock feed. Hester used the advances in commodities and in cash provided by Taylor to feed and pay the laborers who were engaged in cutting, hauling, and rafting timber for him. On one occasion, A. J. Thomas, who was employed by Hester, received from Taylor the sum of $10.40 drawn against the wages due him from Hester. Taylor in turn charged this sum against Hester's future log deliveries. Meanwhile, Hester had been rafting timber to the Taylor and Myers mills. For the logs cut on Taylor's land, Hester was paid forty-one cents each, and for those cut on his own property or on the public domain or state lands, he was paid sixty-one cents each. In the spring of 1861 Hester settled his accounts with Taylor and marched off to war. At that time the balance to his credit above deductions for advances of cash and supplies was $157.74.

Alexander Scarbrough, who lived near Tuxechaena Creek, was one of the earliest logmen in the coast country. He sold logs to Taylor from time to time during the period 1847-1860. During January, 1847, for example, Scarbrough sold 300 logs to Taylor for $137 at the rate of $3 per thousand board feet. Out of this sum Scarbrough paid the costs of felling, hauling, and rafting. The timber, however, was probably cut from the public domain and therefore cost Scarbrough nothing. On another occasion Scarbrough agreed to brand 1,000 logs and put them into the creek. On their part, Taylor and Myers promised to pay thirty cents a log for every batch of 100 logs Scarbrough put into the creek. After the last unit of 100 logs had been delivered to the creek, the logman was promised a bonus of $25.

Business dealings between millmen and logmen are

further illustrated in the contract between Taylor and
Jacob Stronger, dated January 22, 1847. Stronger agreed
to deliver 1,000 logs within a period of four months, the
size of the smaller end of the logs to be not less than eighteen inches. Clearly only the largest and best timber was
being removed from the forest in 1847. In 1859-1860 Hiram
Williams, a small slave owner, rafted logs to the Taylor
mills for 73½ cents each. He was paid twenty cents each for
rafting logs belonging to Taylor. The rates paid Williams
show that the value of logs had increased, but the increase
was much less than that in the price of lumber.

The logging and milling business employed a considerable number of Negro laborers in the late ante-bellum
period. Taylor was a slaveholder when he arrived at
Mississippi City in 1846, and from the beginning he employed slave labor in his mills. In 1846 the Davis-Taylor
firm paid $31 for the hire of Jasper, a Negro slave. The
owners of another slave employed for three months were
paid $49.04. The owner of Fanny, a Negro woman, re-
ceived $100 for one year of her labor. Taylor and Davis
paid $12 per month for the labor of Day, another slave.[12]

By 1856 Taylor had acquired thirteen slaves. Two years
later he purchased five more slaves from the Fowler estate.
For each of the three men Taylor paid $1,200. Apparently
slaves constituted the entire labor force of the Taylor mills
except for the foreman and a few other specialized employees. Five Negroes owned by Hiram Williams and three
others belonging to E. A. Bradford were employed in the
Taylor mills in 1859-1861. The owner received $20 per
month for each slave Taylor hired. The three Bradford
Negroes, on account of illness, lost a total of thirty-two days
during the year 1859. In the same period the slaves of
William Ramsey were employed by Taylor. For each of

his five slaves John Fairley received $18 per month. The slave of William Bond ran away in July, 1858, but returned the next month. Four slaves of Oliver Cowan, a local merchant, were hired by Taylor. In 1860 two slaves owned by John L. Dantzler were employed in the mills. The firm of Taylor and Myers paid out very little cash for the slave labor, since most of the slave owners accepted provisions instead.[13]

It seems evident that Negro slaves, who provided a stable supply of labor, were used profitably in the mills belonging to Taylor and Myers and also by others engaged in forest industries.

The work day in the sawmills extended from dawn to dark. Handling of the heavy, green, unseasoned lumber required great physical strength and endurance. In the coast country, where both humidity and temperatures were high for most of the year, the Negro was perhaps better adapted than the white man to the usual tasks inside the mills. It would seem that the simplicity of tasks such as firing boilers, stacking lumber, and manipulating crude carriage blocks, and the ease with which close supervision could be exercised in the mills, made the employment of slave labor highly advantageous.

Although most of the labor in the mills was performed by Negro slaves, logging and rafting were usually done by white labor. It appears that loggers, timber cutters, and rafters were paid on a piecemeal basis. Log choppers received ten cents a log for felling trees. In one case John Bond cut 250 logs and received $11.22 for his labor after purchases from Taylor had been deducted. Another log chopper, Bargie Bond, made an average wage of $1.00 per day. Rafting required greater skills than chopping and brought higher remuneration. John Brown, a Biloxi River rafter, was paid a daily wage of $1.50. Charles Rushing, a rafting and logging foreman, received $37.20 per month.[14]

The facts indicate that lumber operations in Harrison County were profitable in the last few years before the Civil War. Lumbermen adopted the latest technological improvements in their mills and expanded productive capacity. Even through slaves were generally employed in the mills, specialization of labor developed. These early sawmill operations were significant. They opened up new opportunities for employment of a growing population along the coast and in the back country. They made possible a shift from a pastoral economy that was being undermined by population pressure to new ways of earning a livelihood. In time the new economic system would supplant a pioneer society that had existed almost unchanged for generations.

CHAPTER IV

YEARS OF TRANSITION

T HOUGH THE SOUTHERN LUMBERING INDUSTRY was barely beyond the pioneer stage in 1860, the yellow pine forest industries along the Gulf Coast and the neighboring railroads had experienced a remarkable growth in the last ante-bellum years. By 1860 longleaf pine lumber and logs from Mississippi were being marketed in St. Louis, in cities along the Atlantic seaboard, and in Western Europe.

The extent of the damage inflicted on the longleaf yellow pine industry in Mississippi during the Civil War is difficult to estimate. Some of the smaller mills along the coast were unmolested. In one instance, William Griffin concealed his mill in the marshes above Moss Point to keep it from falling into the hands of the enemy. Less fortunate were the owners of the Goode mills, the sites of which were occupied by detachments of the invading armies; the mills themselves were destroyed. The boilers and foundries of some of the mills in the Bayou Bernard area were appropriated to assist the Confederate war effort. On the other hand, after the capture of New Orleans, a lucrative trade in lumber sprang up between that city and the Mississippi Coast, a trade with which the occupying Federal forces did not interfere.[1]

After the war lumbering was one of the first of the southern industries to revive. In 1865 lumber brought from $24 to $30 per thousand on the New Orleans market, a price beyond the wildest dreams of ante-bellum lumbermen. Consequently, all along the Mississippi Coast smoke began to pour out of sawmill boilers, and again the whine of circular, gang, and sash saws was heard along the banks of the rivers and bayous. This postwar prosperity was short-lived, however. Its source was a flourishing trade in lumber with Mexico. When Maximilian was dethroned and shot, this highly speculative trade ended. Prices of lumber came tumbling down, and many Gulf Coast lumbermen went into bankruptcy.[2]

With the exception of this brief spurt of Mexican-inspired prosperity, the period between 1865 and 1890 was one of slow but steady development of the forest industries in Mississippi, characterized by small manufacturing units and limited capital. It was, moreover, a period of transition, for the improved mechanization typical of American postwar industry in general had, by 1890, completely revolutionized the yellow pine industry. The new techniques of lumber manufacture required a larger aggregation of labor in general and more varied and more highly skilled labor in particular, a larger outlay of capital investment, and more complex forms of business administration. Contributing to the same general result was the postwar acceleration of railroad construction, greatly increasing the consumption of lumber products, affording transportation to markets formerly inaccessible, and opening up new areas of pine lands for exploitation by millmen.

Before the yellow pine industry could reach its maximum development, certain early postwar handicaps had to be removed. Chief of these, perhaps, was the dearth of markets. Pre-Civil War markets, both foreign and domestic, had been for the most part disrupted by war and the accompanying blockade of southern ports. During the war foreign lumber merchants had found new sources of supply, and, after the war, were slow to renew their contracts with the Gulf Coast millmen. As production increased, it became imperative that new markets be found and old ones recaptured. As in other days, New Orleans was the chief outlet for Gulf Coast lumber. By 1870 trade with Texas, Cuba, and Mexico had been revived. Small quantities of lumber were being sent to New York, Boston, and Philadelphia, and such items as spars, masts, and square timber were being shipped across the Atlantic. Nevertheless, prior to 1870 markets were still too limited and lumber too cheap for the speedy expansion of the lumber industry.

In the early postwar years many new mechanical inventions and improvements were introduced into the mills of the Lake states and later brought to the South. The single circular saw which had come into general use in the late forties and early fifties was superseded by a double saw, in which one blade was mounted above the other. This improvement enabled the saws to cut logs any dimension. The speed of the saws was also increased, and in the 1870's the daily capacity of a few of the circular mills was between 40,000 and 60,000 feet, depending upon the size of the mill and grades of lumber sawed.[3]

Gang saws, consisting of a series of adjustable blades set

in a frame, were able to reduce a log into boards in one operation. In many mills where the maximum output was desired, circular saws were used to remove the slabs from the logs, while gang saws were employed to complete the sawing operations. Although the output of the gang saw was less than that of the circular saw, it cut a more uniform product with less waste.

During the decade after 1865 many other remarkable improvements in mill machinery made it possible to increase the output of the mill and also to produce a superior commodity. The introduction of friction feed, wire feed, and steam feed stepped up the speed of the carriage. Labor-saving devices such as head blocks, dogs, and "steam niggers," permitted the logs on the carriage to be handled mechanically and thus eliminated much of the back-breaking work that had been the lot of the ante-bellum sawmill worker. In addition endless chains were introduced to convey logs from pond to mill, lumber from saws to storage ramps, and waste from mill to refuse dumps. In the sixties and early seventies the double-edger and gang-edger appeared in the Lake states, which shortened the time required for finishing boards. By the early eighties, some of the new mechanical improvements had been adopted by a few Mississippi Coast lumbermen.

The first crude dry kilns, erected in Mississippi probably during the late seventies or early eighties, contributed much to the subsequent development of the yellow pine industry. Before the kiln was invented, lumber was cured by air-drying in the open. In the South, where rainfall was heavy, weeks often passed before the lumber became thoroughly dry. The advantage of kiln-drying over air-drying was two-fold. After going through the kiln, the weight of lumber was reduced, and the wood was less subject to bluing, a discoloration caused by the attack of fungi. The weight reduction was especially important since it lowered shipping costs.[4]

Few of the technological developments mentioned above had reached the pine country of Mississippi by 1870. With the exception of one mill at the mouth of the Pearl River, and possibly two in the Pascagoula-Moss Point area, sawmills employed the same techniques as in ante-bellum times. In the Pascagoula district eight mills produced approximately 35,000,000 board feet from May 31, 1869, to June 1, 1870. Of this total the mills owned by Walter Denny and William Griffin accounted for 29,000,000 board feet. The Denny circular sawmills and the Griffin gang mills were forerunners of the large manufacturing units that would ultimately prevail in a country of virgin forests.[5]

Although labor and timber were plentiful and cheap, lumbering in the Pascagoula-Moss Point area was handicapped by a scarcity of shipping and markets. Up to 1870 the market was confined chiefly to Mexico, the Caribbean area, and New Orleans, though small cargoes of lumber had begun to reach the large cities on the Atlantic seaboard and in Europe. In 1874 George Denny of Moss Point received an order for one-half million board feet of lumber from the Amazon River basin, but was delayed in filling the commitment because of a scarcity of ships. With the growing demand for lumber Canadian vessels eventually were shifted to the Gulf ports during the winter months, thus to some extent lessening the transportation shortage. Even with a more adequate supply of ships the Moss Point-Pascagoula millmen were handicapped by a shallow harbor that prevented lumber vessels of 100,000 board feet capacity or more from loading a full cargo at the mill. The ships completed their cargoes nine miles away at Horn Island, where lumber was brought from the mills by rafts and lighters having a capacity of 40,000 feet each.

The expansion of the lumber industry which made the Moss Point-Pascagoula area the second largest lumber center on the Gulf of Mexico in the middle seventies may be measured by the rise in production, which was about 40,000,000 board feet in 1873, and about 50,000,000 board feet in 1877. In 1873 there were ten mills, and in 1874, eighteen. By the end of the decade gang and circular saws had taken the place of most of the old up-and-down and muley saws. During these years an unknown quantity of spars and hewn timber was also brought down from the interior on the waters of the Pascagoula and Escatawpa rivers.

The lumber industry was not greatly injured by the Panic of 1873. Although some lumbermen failed, others entered the business when costs were low. Emile De Smet, a Belgian national with access to foreign capital, for example, acquired a number of mills and for a time was perhaps the largest operator in the coast country. He was known as a "human cyclone" by his contemporaries, mainly because of the rapidity with which he enlarged the scope of his business operations. De Smet was the first to introduce electric lighting in Pascagoula mills for continuous operations around the clock.

To some extent the expansion described above can be attributed to increased prices and the construction of a railroad from Mobile to New Orleans. From 1874 to 1877 lumber prices ranged from $10 to $15 per thousand board feet, depending upon the grade. At the same time timber could be obtained from logmen at the mills for from $3 to $5 per thousand. Square timber hewed in the woods bordering on the tributaries of the Pascagoula sold for eleven cents a cubic foot. With the building of the railroad which later became the Louisville and Nashville, many small mills were erected, and their products came to Pascagoula for overseas shipment. In addition, the railroad, in order to obtain materials to maintain permanent bridges across the bayous, in 1873 erected one of the first wood preservation plants in the United States. In this plant piling was treated in large vats containing creosote and other chemicals which made the piles resistant to the deteriorating effects of the brackish bayou waters. After the turn of the century many such creosote plants were erected in Mississippi.

The depression of 1873 was only a memory in 1877, when the twenty-five mills, mostly small, in the Pascagoula-Moss Point district were operating at maximum capacity and producing around 60,000,000 board feet.[6] This period of prosperity was interrupted when agents of the Department of Justice, believing that the timber supply of the mills was being unlawfully taken from Federal lands, seized all the logs and lumber at most of the mills without warning. For a time not a single schooner was allowed to put out to sea from the Pascagoula Bay area; few of the mills were able to operate continuously between the middle of 1877 and the last quarter of 1879.[7] Nevertheless, during the last three months of 1879 after all the mills had resumed operation, their production was greater than for the past two-year period. In 1880 eleven saw mills were located at Moss Point with a combined daily capacity of 220,000 feet. The same year boards and scantlings for building purposes were sent to Cuba, Mexico, South America, and the Windward Islands. A smaller quantity of deals for shipbuilding went to France, Holland, Belgium, Spain, and Germany from the coast country.[8] (See Map II.)

In the middle eighties James Hunter, born in Scotland,

came to Mobile and established a large export business. Later Robert Hunter, a brother of James, and the Benns formed the Hunter-Benn Lumber Company, a subsidiary of Price and Pierce, international timber merchants, and operated a number of mills on the coasts of Alabama and Mississippi.

Merchants and exporters, believing that yellow pine had a profitable future, provided a considerable amount of capital for mill expansions. This enabled some of the millmen to obtain the latest technological equipment and to make the transition from small to large-scale operations. The W. Denny Company's production of 15,000,000 feet in 1870 had grown to 55,000,000 in the late eighties. Denny's mills, located near the mouth of the Escatawpa River, incorporated many of the latest sawmill machines in use at that time.

A business typical of many in the Moss Point district was that of L. N. Dantzler. In 1873 Dantzler purchased a small gang mill from his father-in-law, William Griffin. In partnership with Evans, Dantzler operated a shipyard, later adding a shingle business, and another sawmill. In 1884 Henry G. Buddig, a New Orleans lumber dealer, suggested to Dantzler that he expand his mill business, and offered to provide the necessary capital. Dantzler accepted the offer and built one of the most modern, well-equipped mills in the coast country. Completed in 1885, the mill cost much more than anticipated, for 2,000 pilings had to be driven for the lumber ramp, and a firm foundation for the heavy mill in the soft marshland required large amounts of cement. The mill was equipped with five boilers, a 500-horsepower engine, double circular and gang saws, edgers, and a complete manufacturing plant including dry kilns, planing, milling, smoothing, and edging machines with which to turn out a finished product. Endless chains conveyed the logs from the lake to the saws and transported the waste materials 300 feet away from the mill buildings. The lumber ramp had a storage capacity for 5,000,000 board feet.[9]

Other lumbermen enlarged the capacity of their mills, and new manufacturing plants were established. Howze and Wyatt Griffin at Moss Point, who operated a small mill in the early seventies, in 1889 erected a plant with a daily capacity of 75,000 feet. Wyatt Griffin, a grandson of William Griffin, in the nineties became the sole owner of the Moss Point Lumber Company. Will Farnsworth, financed by the Hunter-Benn Company, became the manager and part owner of a large sawmill at Pascagoula. Following De Smet's failure, due to overexpansion and a slight depression in the lumber business, George Robinson acquired the De Smet properties in 1883 and added the latest improvements to his mills. New improvements which enlarged mill capacity and increased the efficiency of sawing came into general use in the Pascagoula district during the years 1884-1890.

The years 1885-1890 were a boom period for lumbermen. Sawmills, with more orders than could be filled, often operated both night and day. With the erection of large manufacturing units a number of related industries such as foundries and machine shops sprang up to serve the lumber industry. This period of high output is illustrated in the available production figures for the Pascagoula area from 1879-1880 to 1891-1892.[10]

YEAR	BOARD FEET
1879-1880	60,000,000
1883-1884	67,308,000
1884-1885	67,839,000
1885-1886	—
1886-1887	70,000,000

1887-1888	—
1888-1889	107,000,000
1889-1890	119,255,000
1890-1891	170,000,000
1891-1892	127,002,000

The boom in the lumber business was also evidenced by the large rafts of timber and the thousands of loose logs that came down the river in times of high water. In Pascagoula Bay, ships of many nations could be seen making their way slowly up the channels to Moss Point for cargoes of lumber. In 1890 there were four large mills at Moss Point with a total daily capacity of 384,000 board feet, and the town's 4,000 inhabitants all depended upon lumbering for a livelihood. By this date logging and rafting had all but displaced the pioneer economy in the interior.

The lumber industry at the mouth of the Pearl River revived quickly with the end of the Civil War; in fact, there is evidence to suggest that some of the mills operated during the war years with little interruption. The mill business owned by Joseph Favre and W. J. Poitevent, who were brothers-in-law, was the largest in Mississippi in 1870. They employed 155 workers and the annual production of their mills was around 20,000,000 feet. In 1872 the Poitevent-Favre Lumber Company supplied the lumber, ties, and piling for the construction of the bridges erected by the Mobile and New Orleans Railroad Company. For furnishing the timber for the construction of the jetties at the mouth of the Mississippi, Captain Poitevent was styled by his contemporaries the "Lumber King of the South."

Poitevent and Favre owned three mills and a shipyard at Pearlington and operated a line of steamers and schooners in the New Orleans and Caribbean lumber trade. This company was perhaps the first in Mississippi to adopt the technical improvements which appeared in the sawmills of the Lake states in the early seventies. In 1879 or 1880 Dr. Charles Mohr stated that the new Poitevent and Favre mills had a potential daily capacity of 100,000 board feet.[11] Ten years later the company was cutting over 30,000,000 feet yearly and was one of the largest producers of lumber in the South. The firm also employed eighteen vessels in the lumber trade.

Up to the late seventies most of the timber for the Poitevent mills had come from the forests located on the Pearl River and its tributaries. But the passing of the timberlands into private ownership forced the company to turn elsewhere for supplies of raw materials. Large areas of forest lands were acquired in St. Tammany Parish, Louisiana, and by 1884 a standard gauge railroad known as the East Louisiana Railroad was constructed to the timberlands. Logs were rolled into West Pearl River at Floransville and from there drifted down to the company boom located one mile from Pearlington. They were then towed to the mills through Jug Bayou and the half-mile long, log-ribbed canal built in 1879 that connected East Pearl River to the bayou.[12]

Another ante-bellum lumber company, the Weston-Carré partnership, resumed operations soon after the war. Their sawmill business, the second largest in the Pearl River district, produced 5,110,000 feet of lumber in 1870. Planing mills were added in 1872. In 1874, when the partnership was dissolved, the company owned two mills: a circular mill at Bogahoma Creek and another at Logtown. Henry Weston acquired the Logtown mill. Expansion of the Weston Company came in 1889 when improvements increased the output of the mill. Henry Weston, the founder of the company, was past the prime of life, but his sons who had grown up in the sawmills assumed most of the responsibility and direction of the mill business during the late eighties. The younger Westons provided leadership at a time when the lumber industry was entering a period of vigorous growth, and the company expanded rapidly.[13]

In 1873 there were twelve sawmills near the mouth of the Pearl River, and six others a few miles to the east in the Hancock County area of the Bay of St. Louis near the mouth of the Wolf and Jordan rivers. Three years later eight hundred people were reported to be dependent for livelihood on the large mills at Pearlington and Logtown. Besides the lumber shipped from the Hancock County mills, spar timber was sent to the markets outside.

As in the Pascagoula district, the Pearl River mills were forced for a time to suspend operations in 1877 when, as mentioned earlier, Federal agents seized both logs and lumber at the mills. This action by the Government resulted from the belief that the timber sawn by Pearl River mills came from Federal lands.

In the early eighties timber from all the tributaries of the lower Pearl, Pascagoula, and other coastal streams was being rafted to the coast mills. One observer reported that people who lived in the Pearl River regions had abandoned their agricultural pursuits to become timber workers. Logtown and Pearlington, together with Moss Point in the Pascagoula area and Handsboro on Bayou Bernard may properly be regarded as the first four Mississippi towns that grew directly out of the lumber industry. Although lumber output in the Pearl River district advanced in the years 1880-1890, the rate of growth was much less than on the Pascagoula, largely because the timber areas accessible to water transportation were more limited and interior mills were being erected in the middle eighties along the newly constructed New Orleans and Northeastern Railroad, which bisected the Pearl River district.

After the Civil War the size and scope of the lumber industry in the Bayou Bernard district and the Bay of St. Louis area remained relatively insignificant in comparison with the Pearl and Pascagoula areas. In 1870 the production of the Denny mills at Moss Point was about equal to the total output of all of the mills in Harrison County. Calvin Taylor continued his ante-bellum lumber business, but was handicapped by a lack of capital and unable to make the transition to large-scale operations required of those who would succeed in the postwar industry. His plight was common to almost all the ante-bellum sawmill men who remained in the Bayou Bernard area. Taylor continued in business until his death in 1886, but for the entire postwar period his average production was less than it had been in 1860.

Besides scarcity of capital, the absence of prime timber near the rafting streams was a handicap to the Bayou Bernard millmen. Henry Leinhard was by reason of ability and conservative management one of the few who succeeded. In 1867 Leinhard, who had entered the lumber business about 1858, built what may have been the first crude tramroad for transporting logs in Mississippi. It bore little resemblance to the heavy steel rails that were to be used for later logging operations, for it was constructed of wooden scantlings placed end to end to form the tracks upon which trucks containing only a few hundred feet of logs were pulled by mules and oxen. Later after Leinhard's mill business grew, a pole road was constructed by laying poles end to end to form rails, and a steam locomotive weighing from five to ten tons was used to pull the log cars.[14]

When prices of lumber were low, manufacturing was unprofitable without the cheap freight rates offered by schooners and brigs. In the early period when only superior trees were utilized, rafting of logs was cheaper than other means of transportation, but when at a later date only small timber remained, losses through sinkage made water transportation uneconomical. The advantages of water transportation had enabled a few coast lumbermen to begin lumber manufacturing when capital requirements were low, and these individuals had been able to increase the size of their operations along with an expanding industry. With the construction of railroads and the development of domestic markets during the 1880's, however, the center of lumber manufacturing shifted from the Gulf Coast into the interior.

CHAPTER V

COMMERCIAL LUMBERING
IN THE INTERIOR, 1865-1890

APPARENTLY THE SMALL ANTE-BELLUM LUMBER INDUSTRY in the interior, especially that located in the path of military operations, was almost totally destroyed during the Civil War. The markets that these mills served, primarily the small towns and plantations near the railroads, were likewise demoralized by the war. Postwar shipments to places north of the Ohio River, if made at all, were of little consequence. As late as 1875 yellow pine remained almost unknown in the great consuming areas of the North. Moreover, the white pine industry in the Northwest was reaching its peak, and prices of lumber in Chicago were little more than the freight costs on yellow pine from south Mississippi to Chicago.[1]

In the late sixties steam circular sawmills with an average capacity of 10,000 feet or less sprang up along the Illinois Central Railroad and the Mobile and Ohio. These mills were generally erected in the forests. Their lumber was transported to the railroads on wooden tracks pulled by ox and mule teams. Their markets, limited mainly to New Orleans and to small towns near the railroads in north Mississippi and Tennessee, were inadequate to stimulate a rapid expansion of the industry. At Summit, Mississippi, J. J. White, a pioneer lumberman, was running a small sawmill. About 1868 James Haskins erected what was considered "a first class mill" west of the Pearl River near Brookhaven.

The early postwar state of lumbering in the interior is reflected by the United States Census of 1870. The eleven small steam mills in Pike County were about the same as in 1860, as were the few mills on the Mobile and Ohio Railroad in Wayne and Clarke counties. The Variety Works in Copiah County, producing 10,000,000 board feet in the census year 1870, was a large operation for the back country. At Fruitdale on the line of the Mobile and Ohio, one steam mill cut dimension stock, presumably for the export trade. Nevertheless, lumbering throughout most of the back country was less developed in 1870 than in 1860. Indeed, Lawrence County had fewer mills in 1870 than in 1840.[2]

Expansion of the yellow pine industry in the interior of Mississippi came with the development of a market for railroad car sills. In the early seventies car manufacturing companies began the experiment of using yellow pine in limited amounts to replace such woods as oak and white pine. The experiment succeeded; yellow pine not only possessed the exceptional strength necessary for car manufacture, but also sold for less than other varieties.

In the years 1873-1880 the car sill market grew. The pioneer buyers of yellow pine were the Lafayette Car Shops, Indianapolis Car Foundry Company, Ohio Falls Car Company, and the Missouri Car Foundry Company. These companies paid $15 to $16 per thousand board feet for sills. The price and market stimulated the development of lumbering on the Illinois Central. Many circular mills cutting 10,000 to 12,000 board feet were built primarily for the purpose of sawing car materials.

In the late seventies yellow pine began to compete with white pine in the North on a small scale. Orders for yellow pine lumber were numerous, but only the better grades of lumber were acceptable to northern buyers. Sideboard strips also were purchased by northern buyers and converted into ceiling and flooring. But much of the southern lumber was air dried and heavy, and for this reason could not bear the transportation costs.

In this early period manufacturers of yellow pine were compelled to keep the prices of their product low in order to compete with white pine in the northern market. They were further handicapped by having to dispose of their products through the medium of commission men, thus retaining little control over the sales price of their lumber. The adoption of the dry kiln eventually gave yellow pine manufacturers access to markets north of the Ohio River. Kiln drying produced a superior grade of lumber and reduced the weight of a carload from 40,000 to 23,000 pounds. The average weight of a thousand board feet of rough lumber was 4,300 pounds; of dry kilned dressed lumber, 3,100 pounds.[3]

Even with the improved manufacturing techniques, yellow pine lumber was slow to gain favor with northern consumers. When it was first introduced, it encountered the innate conservatism of the average man who distrusts that which is unfamiliar. Perhaps the prejudices of carpenters were the most serious barrier; accustomed to the softness and ease with which white pine could be shaped and worked, they found yellow pine difficult to handle.

Others handicaps to the sale of yellow pine in the North were the beliefs that it would not take paint and that it decayed rapidly. These mistaken notions, kept alive by white pine merchants and others who feared the competition of southern pine, took years to erase.

T. K. Edwards, freight agent of the Illinois Central, has been given credit for sending the first carload of yellow pine lumber north of the Ohio River in the late seventies. It seems that Edwards without having a buyer sent the lumber to be sold at whatever price it would bring. J. J. White is reported to have sent to Chicago one of the first carloads of lumber from the mills located on the Illinois Central Railroad. W. C. Ott also was one of the first to offer yellow pine for sale in the Chicago market.

F. W. Norwood, a shoe salesman who became interested in yellow pine lumber, about 1884 established one of the first retail lumber yards to handle southern pine lumber in Chicago. He purchased lumber from Haskins and Hamilton, who owned a dry kiln and planing mill at Brookhaven.

In 1884 or 1885 C. S. Butterfield joined Norwood's firm, and subsequently the partners contracted to furnish 2,000,000 board feet of southern lumber for the construction of the Marshall Field warehouse at Chicago. To meet their obligations the Norwood-Butterfield firm contracted for the entire cut of several small mills located near Brookhaven. Because of the inefficiency of the mills, however, the lumber failed to meet specifications. Under threat of heavy penalties for failure to fulfill their contract, the partners came south to Brookhaven and purchased several small sawmills, a planing mill, and a dry kiln. Their business expanded, and in 1890 they were among the largest yellow pine manufacturers in Mississippi.

From a small beginning in 1869, lumber by 1885 had become the largest single class of freight carried by the Illinois Central Railroad. Yellow pine, though still relatively unknown in the great consuming areas of the country, increased in importance with the growing shortage of white pine timber. In addition, the fantastic expansion of American industry in this era brought a growing demand for large supplies of lumber. Yellow pine was cheap, selling in some grades for $13 per thousand feet less than white pine in the Great Plains region. As a result southern pine was being sold in Omaha, Topeka, and Springfield, Missouri, in 1886-1887.

The expansion of the yellow pine industry was greatly aided by the dwindling supplies of white pine, not only because the market for yellow pine improved, but also because capital amassed in the white pine industry now began to flow into the development of the yellow pine industry. The existence of millions of acres of cheap virgin timberlands fired the imagination of speculators, who realized that again, as in the Lake states, large profits could be gained by buying lands and holding them for an increase in stumpage values.

Newspapers and lumber journals did much to publicize the young growing lumber industry in the South. The editor of the St. Louis *Press*, after making two trips to the pine country of south Mississippi in 1874, reported in amazement that the forests were almost untouched. He considered that lumbering opportunities were unlimited in this area, for land in large tracts situated close to rafting streams could be purchased for $2 an acre.[4] The editor of the *Northwestern Lumberman* wrote in 1880 that similar conditions existed on the Pearl River. Another northern lumberman after computing the cost of production in the

South concluded in 1877 that a profit of $7 per thousand board feet was possible by shipping lumber to the New York market.

Although lumber journals were pointing out the advantages of investing in the southern yellow pine industry, northern capitalists hesitated nevertheless. Some lumbermen whose conception of the South was derived from the propaganda of radical Republican politicians believed that both life and property were unsafe in the Southern states. Seeking to overcome this fear, the editor of the *Northwestern Lumberman* assured lumbermen that property was as secure in the South as in Chicago, and he predicted that the South was destined to become the most important field for lumbering on the continent. Its timber resources were vast and potential markets for southern lumber existed in all parts of the world. The great drawback, according to this editor, was the want of capital for development. Although advantages for profitable investment were great, logging still was more expensive in the South than in the North because both labor and livestock were inferior to those of the North.

In the eighties expansion of southern lumbering resulted from the broadening of markets rather than an influx of northern capital, for the first northern investments were limited chiefly to the purchase of timberlands. In this period large purchases of land were made by northern speculators who hoped to realize a profit from the increase of timber values. In 1882-1883, for example, the woods were said to be full of northern prospectors. Indeed, the purchase by land speculators of lands near lines of transportation may have slowed the growth of lumbering, because they held the timber off the market in expectation of a rise in stumpage prices. Some of the small millmen in the late eighties, when lumber prices were extremely low, could not meet the speculator's price and had to cease operation.

In comparison with its slow development during the seventies, the lumber industry on the Illinois Central Railroad expanded rapidly in the eighties. In 1883 it was reported that there was a sawmill at every train stop of the railroad, and J. J. White (1830-1912), starting with a small mill shortly before the Civil War at Summit and another in 1873 at McComb, was the "first man to build a tramroad in the state of Mississippi to be used for hauling logs to the sawmill by a steam locomotive." White's mill business expanded over the years to become one of the most prosperous operations in the state.

Isaac C. Enochs, a native of Crystal Springs, Mississippi, was another pioneer lumberman operating on the route of the Illinois Central Railroad. As a young man Enochs managed a portable mill owned by his father with intention of acquiring the funds necessary to attend law school. On the day he became proprietor of the mill, the town of Crystal Springs burned, and after the fire the mill boiler exploded, destroying the plant. Enochs replaced the mill at Crystal Springs and erected in the Fernwood vicinity two other mills which had a combined daily output of 40,000 board feet. His first logging road was constructed of wood and the log trucks were pulled by mules.[5] Later a locomotive that moved over wooden rails was acquired. In 1884 the firm erected a sash and blind factory at Jackson, Mississippi. He and his brothers operated the Fernwood mills and a number of portable mills. By 1890 the Enochses had expanded mill operations and acquired large holdings of timberland.[6]

By the mid-eighties a wide swath had been cut in the

forest along both sides of the Illinois Central Railroad all the way from northern Copiah County to the marshes immediately north of Lake Pontchartrain. Lumber in ever-increasing quantities was reaching northern cities from the southern mills. Millmen with sufficient capital were expanding the scope of their operations by increasing mill capacity, building tramroads, and acquiring additional timberlands. Failure was to be the lot of those whose financial resources were inadequate to meet the new conditions necessary for successful lumber manufacture.

Besides the new capital requirements for mill expansion, the removal of most of the timber from near the railroads forced lumbermen to invest large sums in construction of log roads. One of the earliest types, the Cole pole road, served as a makeshift until a more elaborate type could be built. Invented by an Alabamian, John Cole, and first used by Henry Watson in west Florida, it was constructed of round poles twelve inches in diameter at the larger end and eight inches at the smaller laid end to end to form track. Traction came from a locomotive with wide wheels deeply grooved to fit the pole track. Cost of the road, between $150 and $200 a mile, was much less than for roads of heavy iron rails, but its hauling capacity was low, limited to about 300 feet of logs per car. When larger mills were erected, conventional roads, both narrow and broad gauge, were built to provide the mills with raw materials. In 1887 fifteen millmen on the Illinois Central were operating tramways, a few of which subsequently developed into common carriers. (See Plate I, Figure 1.)

By 1887 a number of small lumber and distribution centers had grown up along the line of the Illinois Central. Each week during the year around 300 carloads of lumber were shipped northward. The production of the seventeen mills in the immediate vicinity of Brookhaven in 1886 was estimated at 18,000,000 board feet monthly. In 1888 there were seventy-five mills within a thirty-mile radius of McComb. The principal shipping points were Ironsides, McComb, Summit, Bogue Chitto, Wesson, Brookhaven, Hazlehurst, Johnston Station, and Crystal Springs.[7]

Lumbering developed at a slower pace on the Mobile and Ohio Railroad, primarily because the railroad leased large tracts of timberland, approximately 750,000 acres, to A. C. Danner, who then subleased the land to turpentine operators.[8] Another handicap arose when the state and Federal lands passed into the hands of large land syndicates. Much of the territory traversed by the Mobile and Ohio was drained by the Pascagoula and its tributaries, and in the vicinity of the railroad lumbering consisted chiefly of rafting logs to Moss Point.[9]

In the middle eighties a vast new area of pine country was opened up for exploitation by the construction of the New Orleans and Northeastern Railroad from New Orleans to Meridian, Mississippi. It was the first road to be built through the large virgin pine woods between the Pearl and Pascagoula rivers. Although small amounts of timber had been removed from the areas bordering on rafting streams, and much of the spar timber had long since disappeared, most of the magnificent timberlands remained in their original state when the railroad was completed.

A few of the small towns along the New Orleans and Northeastern Railroad became lumber manufacturing and distributing centers. One was Lumberton which had its beginning in 1886 when the Hinton brothers, John, Robert, and H. P., and their cousin, Herbert Camp, erected a mill there. John Hinton had drifted to McComb, Mississippi, from Georgia in 1878, and found employment in the J. J.

White Lumber Company mill. After a few years Hinton
had become general superintendent of the White mill
and, perhaps of more importance for his future, had married the daughter of his employer. The Hintons and Camp
acquired timberlands for $1.25 an acre near what was to
become Lumberton. In 1886 the company owned planing
machines and a dry kiln besides the sawmill.[10]

A second railroad destined to be of even greater importance to the development of forest industries was the Gulf
and Ship Island. Its route ran between the Pearl and the
Pascagoula rivers all the way through the pine country from
Jackson to the Gulf of Mexico. The original Gulf and Ship
Island Company had received a large grant of Federal
land in 1856, but the Civil War and subsequent scarcity of
capital had delayed the beginning of construction. In the
early eighties a new company began building a line, but
failed. Another company headed by Joseph T. Jones, a
wealthy Pennsylvania capitalist, completed the road to
Jackson in 1901. The south Mississippi areas through which
the road would pass had changed little since De Soto's
time. A writer in the Biloxi *Herald*, February 18, 1888,
observed:

> The pine forests are a howling wilderness in some of the
> counties of South Mississippi, and in a week's ride on horse-
> back, a man would see [only] at long intervals a human habi-
> tation. . . . It is a beautiful country with pure air, good clear
> water, and magnificent pines, but without population.

Another writer stated that the people a few miles north of
the coast had changed little from the primitive life of their
forefathers. There were no household conveniences; the
women, using old-fashioned spinning wheels and cards,
made most of the clothing and quilts.[11]

By 1890 the forest industries in the interior of Mississippi were on the threshold of a new era of expansion. The pioneer phase of exploitation of the longleaf pine forests was fast coming to a close. Henceforth large scale lumbering was to prevail in the longleaf pine country of Mississippi.

CHAPTER VI

FEDERAL LAND POLICIES
IN THE PINELANDS

AT THE TIME MISSISSIPPI BECAME A STATE most of the pine forests of south Mississippi were on public lands belonging to the United States Government, and most of them remained Federal property for several decades. Not until the fifties were great areas donated to the states, and very few sales of timberlands were made to private buyers in the meantime.

In 1850, there were approximately 6,000,000 acres of vacant land in the longleaf pine counties of Mississippi. During the period 1811-1847, 9,628,675.51 acres had been offered at public sale and only 950,713.04 acres were sold in the Augusta land district. During the same period, of the 3,466,880 acres in the Washington land district offered at public sale, 1,293,643.65 acres were sold.[1]

Agricultural immigrants generally avoided the heavily timbered pine lands because of the sterility of their soils. Not being farmers, the people who did establish homes and small settlements in the forested country were little concerned with the acquisition of large landholdings. Their pastoral economy required little farmland, though it depended upon vast areas of open range for livestock. Since forest industries were almost unknown except in a few localities, the demand for land in the timbered country at the usual government price of $1.25 an acre was small. Sales in the year July 1, 1850, to June 30, 1851 were only 5,629.77 acres in the Augusta land district and 19,014.97 acres at the land office located at Washington, Mississippi. In the fiscal year July, 1852, to June 30, 1853, 7,814.75 acres were sold at Augusta and 40,259.70 acres in the Washington district.[2]

In the years 1850-1860 small forest industries were expanding. Cheaper timberlands in limited quantities were obtained from the United States Government under the Choctaw Land Donation after 1850, and from the grants derived from the Soldiers and Sailors Act. Donald McBean, Thomas and Rufus Rhodes, Garland Goode, and other lumbermen selected numerous 160-acre tracts with military land warrants and Choctaw land certificates. But the demand for timber still had no great influence on the sales of government timberlands, mainly because state lands— a subject to be dealt with mainly in the next chapter— were being sold at prices much below the minimum price for Federal lands. In addition, sizeable tracts of privately owned land were at the disposal of the lumbermen.[3]

For a number of years after the Civil War, Federal land policy in regard to the South differed from that in other sections of the country. Private entry of land in the South was abolished in 1886. Under the Homestead Act of 1862 a homesteader could obtain 160 acres of land at little cost. But the Land Act of 1866 restricted the amount of land that could be acquired by homesteading in the South to eighty acres in a two-year period. Apparently the Federal Government was seeking to aid the recently emancipated Negroes to obtain small farms at small cost.[4]

During the next few years opposition to this Federal land policy emerged. Critics of the Act of 1866 pointed out that no such restrictions were applied to the disposal of public land in other sections. They also noted that the freedmen were not benefiting as intended because most of the remaining land owned by the Federal Government in the South was "notoriously barren and valuable for only its timber."

Republican Senator James L. Alcorn of Mississippi advocated the enactment of a private entry law allowing anyone who had the money to purchase land in unlimited amounts. He believed that Federal lands in the South would be purchased by those who desired to speculate on the timber if the land were opened to private entry. He argued that the Government would receive nothing for its property if the existing system were continued.

According to Alcorn, the lumber interest which had grown up since the Civil War had stripped timber illegally from hundreds of acres of government lands along the Pearl River and its tributaries. While governor of the state Alcorn had brought these practices to the attention of Federal authorities, but depredations had been allowed to continue unchecked. As the Government was not disposed to guard its timberlands, Alcorn maintained that the public lands ought to be sold. Under a private entry law only about 300,000 or 400,000 acres contiguous to or bordering on the tributaries of streams flowing into the Gulf of Mexico would be sold. The remaining 3,000,000 or more acres of inaccessible pinelands would not bring one cent an acre.

Like Alcorn, Senator Powell Clayton of Arkansas considered the policy of reserving timberland for an agricultural population a mistake. He stated that the Negro had gained little benefit from the Homestead Act:

Your policy would be very humane to the colored man if you were offering him something that was worth anything to him; but you offer him poor, barren lands that have pine timber upon them. In the first place, he must go to work and clear off that pine timber and spend his hard earnings for that purpose; and after he has cleared off the pine timber a partridge cannot live upon the land. . . . The poor men of the South have not made homesteads to any extent upon these lands simply because they are worthless for that purpose.

According to Clayton the homestead laws were perverted by those interested only in obtaining timber cheaply. By submitting an application and paying $5, bogus homesteaders gained the privilege of occupying 160 acres of land for a period of five years. They cut the timber and abandoned the land to the Government.

Mississippi Congressman Hernando De Soto Money stated that the remaining public lands in the South were valuable chiefly for their timber; they lay as a dead weight upon the states and would do so until the crack of doom.[5]

A bill legalizing private sales of public lands was enacted in 1876 at a time when lumbermen and speculators were becoming aware of the potential value of southern yellow pine forests. This law, especially favorable to speculators, placed no limit upon the size of purchases. Under its provisions, land was offered to the highest bidder at public auction, but if there were no buyers the land could be bought later at the minimum price of $1.25 per acre. Thus the minimum government price became the going value of virgin timberlands. In the next twelve years most of the large tracts of pineland passed into private ownership at bargain rates.[6]

The supporters of the 1876 private entry law hoped that thefts of timber on government lands might decrease if lumbermen were allowed to buy them. The problem of such depredations was most difficult to deal with. Logmen, spar getters, coal burners, and naval stores operators for a long time had taken timber with impunity from the national domain. State and Federal authorities seeking to protect public forest lands rarely secured the co-operation of the people of the pine country, who looked upon the timber as being as free as the air. To take a tree growing on government land was not considered a criminal act, as timber had little value, and for most of the nineteenth century the pine forest constituted an obstacle to clearing land for farms. Hence, the average piney woods settler was sympathetic to the lumberman. If his aid were solicited by the Government in an attempt to end theft, he would be strongly antagonistic.

As early as 1817 specific steps were taken to protect timber on land set aside for military purposes. Congress in 1831 passed an act with later amendments providing for a fine of not less than triple the value of timber unlawfully cut on naval reserve timberlands and imprisonment of the culprit not to exceed twelve months. The laws provided for forfeiture of vessel and tackle and a fine for the captain for taking illegally cut timber aboard ship. In 1850 the United States Supreme Court in the case of *The United States vs. Ephraim Briggs* held that the Act of 1831 and subsequent amendments applied not only to timber reserved for naval purposes but also to other trees cut on public lands. Influenced by the Court's decision, government authorities appointed special agents to deal with trespass.

As forest industries expanded the problem of ending theft became more complex and difficult to solve. In 1854 the Land Commissioner of the United States decided to enforce more strictly the laws against trespassing on the public domain. Timber agents were instructed to seize all logs, floating timber, and lumber which had been taken from Federal lands. Depredators apprehended by the agents were not to be allowed to retain possession of stolen logs after paying for them. To retain possession of the stolen logs, trespassers could only buy the land from which the timber had been taken. Payment of a fee in advance for permission to cut timber was prohibited. Where legal process existed, seizure of the stolen logs was to be made by United States marshals and a criminal suit was to be instituted against the guilty persons.

Apparently the policy proved to be unworkable, for in 1855 the system of using special timber agents was discontinued and the responsibility of guarding the timberlands was made a part of the regular duties of district land office registrars and receivers. The land office agents were instructed to extract from deliberate trespassers the value of logs and sell them at public auction. In 1860 compromises with depredators were authorized by the Secretary of the Interior on the following grounds: ". . . entry of land upon which the timber was cut, payment of fifty cents" a thousand board feet stumpage, and costs for expenses incurred in seizure. Later in instances where timber was cut on land not subject to private entry the trespasser was required to pay a reasonable stumpage price according to the market price of the timber—in no case less than a minimum of $2.50 per thousand board feet and costs. The practice of collecting stumpage from depredators became universal in all parts of the country.

In 1877 Carl Schurz was appointed Secretary of the Interior. An immigrant from Germany, Schurz, imbued with the philosophy of forest protection that had prevailed in his fatherland, was alarmed at the waste and rapid disappearance of the virgin forests in the United States. He and his assistant, Joseph Williamson, the Federal Land Commissioner, agreed on the necessity of preserving the forests and of ending the theft of timber from lands owned by the Federal Government. Williamson was amazed to learn that the Government had collected only $199,998.50 at a cost of $45,624.76 during the twenty-year period of compromise with depredators. With the consent of his chief, Williamson relieved the land officers and receivers of their timber duties. He divided the public timberlands into districts and assigned special agents to each. A new policy of strict enforcement of the trespass laws was adopted. Compromises were not to be permitted without approval of the Land Department. Legal proceedings were to be initiated to punish trespassers, or to collect damages for the waste already committed.[7]

Murray A. Carter was appointed agent over the district that included Mississippi, Alabama, Florida, and Louisiana. Soon after assuming his duties Carter informed Williamson of widespread depredations in all of the Southern states. In the coast country of Mississippi agents of the Government seized logs, lumber, and naval stores at the mouth of the Pascagoula and elsewhere. To prevent shipment of forest products from Moss Point and other mills on the river by boat, a Federal marshal armed with a double-barreled shotgun was stationed temporarily on the drawbridge that formed a section of the long railroad spanning the entrance to East Pascagoula Bay. The blockade established by the Government at the mouth of the river made the shipment of forest products by boat to outside markets impossible.

The sudden seizure of logs, lumber, and naval stores by agents of the Government effectively stopped all activities connected with forest industries. Thirty-seven Mississippi Coast sawmills were closed down for a long period, and suffering and hardship were the lot of those whose livelihood depended on employment in the mills and woods. One unnamed observer predicted in late 1877 that many people would soon be on the verge of starvation if the mills were not soon reopened. Another person reported that hunger riots were not far away and that he had actually seen cases of starvation among the unfortunate mill employees. The editor of the Pascagoula *Democrat Star* stated that the seizures had paralyzed the coast industry and that it would take many years for the lumber industry to recover.

At Moss Point the state contested the seizures made at the Griffin mill by the Federal Government, claiming that the logs were cut on lands owned by the state. A posse headed by the sheriff of Jackson County was accused of using force to take the Griffin logs held by the Federal agents. On November 30, 1877, a mass meeting of more than four hundred citizens of Jackson, Perry, and Hancock counties assembled at Scranton, now Pascagoula, to formulate a plan to resist the seizures. A resolution was drawn up which branded as false the assertion of the United States marshal that he and his deputies had been driven away from the Griffin mill by the sheriff and a posse of Jackson County citizens. In another resolution the Federal authorities were charged with a violation of the use of a writ of sequestration in seizing logs and lumber. A petition adopted at the meeting requested an investigation by

Congress. The petitioners stated that as shipowners and as citizens of the United States, Mississippi, Alabama, and Louisiana, their business had been unwarrantedly interfered with by illegal seizures of logs and lumber. According to the petitioners all activities connected with the lumber business had been stopped, and many people were on the verge of starvation. They denied violation of the law and asserted that less than one-tenth of the logs involved had been cut on government lands.

An article in a Mobile newspaper reprinted in the Pascagoula *Democrat Star* declared that over 50,000,000 feet of timber annually had been taken illegally from the national domain in the Southern states. This was vigorously denied by the editor of the Pascagoula *Democrat Star*. According to the Mobile paper a considerable percentage of timber theft had occurred in Mississippi, log seizures in that state had been resisted by armed force, and back country people were preparing to seize the logs and lumber in the custody of the Federal agents. Another writer predicted that there would be war on the Pascagoula. Although local sentiment in the coast country was decidedly opposed to government seizures, it is unlikely that many ever contemplated the use of force. There were more effective and less dangerous weapons with which to deal with the Government.

The cause of the lumbermen whose timber had been seized was helped by the United States district attorney, Luke Lea, who ruled that the mode of procedure used by Federal agents had been illegal and that they had possessed no authority to block the navigation of the Pascagoula River unless absolutely necessary to seize and hold property belonging to the Government. Lea apparently considered the indictments against the alleged depredators unjustified.

His attitude incurred a thinly disguised rebuke from the United States Attorney General, who ordered his assistant to exercise more energy in handling the log cases in the future.

As a result of Lea's ruling, the Government was compelled to release all property seized improperly and subsequently take custody of logs, lumber, and naval stores only under a writ of replevin. Security offered by millmen for possession of the property seized by Federal authorities was unacceptable to the marshal, as each millman volunteered to go on the bond of another in return for a similar service. Later millmen who were given a chance to regain possession of their property by posting bond neglected to do so.

The refusal or failure of the millmen and others to give bond brought forest industries to a standstill for lack of raw material. Nearly all of the logmen, millmen, and naval stores operators were indicted by the Federal Government, and their cases were scheduled for trial in the United States District Court at Jackson during May, 1878. The attorneys of the indicted lumbermen requested an early trial in March, 1878, and attempted to transfer the court session to Scranton on the Gulf Coast. The reasons behind these moves were obvious. At Scranton, where everyone was either directly or indirectly connected with the lumber business, sentiment strongly favored the millmen. Furthermore, an early trial would permit the mills to resume operation at an early date.

In January, 1878, the legislature of Mississippi sent to Congress a memorial requesting speedy relief for those who were dependent upon the log business for their livelihood. The memorial asked that an early trial for the defendants be held at Scranton, and also asked compensation for those persons whose property had been seized.

On February 28, 1878, House Bill 3072 providing for a special term of Federal court to be held at Scranton was introduced in Congress. Schurz and Williamson were strongly opposed to the measure. Schurz regarded the bill as a means of circumventing the Government's cases against the forty-nine defendants who had been accused of taking timber unlawfully from public domain. On the other hand, L. Q. C. Lamar, Senator from Mississippi, led the fight for the court bill. He contended that the methods employed by Federal agents were "harsh, precipitate, and detrimental" to industries in Mississippi. According to Lamar, the lumber trade of an entire section of the state was broken up by the seizure of logs, lumber, and naval stores, and fifteen hundred people "were reduced to destitution" in consequence.

Senator John T. Morgan of Alabama accused the Federal agents first of illegal procedure when they used a writ of sequestration, then later of improper use of the writ of replevin which was the legal procedure in Mississippi courts. Writs issued by the Government, Morgan said, were based upon unsubstantiated testimony and therefore unjustified.

The bill passed both houses easily although a large percentage of congressmen and senators abstained from voting. As President Hayes vetoed the bill, however, the matter was settled contrary to the wishes of the Mississippians.

Despite the failure of the bill to become law, the cases against the lumbermen were not pressed. Indeed, by 1879 almost all of the indictments had either been compromised

or dismissed. Cases against the Pearl River lumbermen
Poitevent and Favre and the H. Weston Lumber Company
were compromised. In the Pascagoula area, however,
cases against the W. J. Griffin and the W. Denny companies were continued. Of all the indictments handed
down, seventy-six in number, only twelve cases remained
to be tried in 1880. These cases were settled by an act
of Congress in 1880 which quashed all Federal indictments for timber theft made prior to March 1, 1879. This
law provided that the accused must purchase the lands
from which they had taken the timber at the minimum
price of $1.25 an acre. In other words, Congress revived
the principle of compromise that had prevailed between
1855 and 1875.

Public opinion was largely responsible for defeating
control of depredations in public lands. So widespread and
of such long standing was the practice of spoliation that
it had become deeply ingrained in the habits of the people.
In the pine country near the Gulf Coast where streams
were numerous a small army would have been required
to prevent theft of timber. Once the stolen timber had
been rolled into the stream, proof of ownership was difficult
to establish. Moreover, the mill owners were not interested
in the origin of the logs they purchased from logmen. In
the southern part of Mississippi government timber had
been taken wherever it was profitable. Indeed by 1880
almost all of the valuable timber had been cut from the
banks of all rafting streams flowing into the Gulf. On
Hobolochitto, Red, and Black creeks 97,116 logs were
reported to have been removed. Also large acreages of
public lands were boxed by turpentine operators.[8]

Mississippi senators who favored reopening of the government timberlands to private entry had maintained that

large-scale depredations would cease once millmen were able to buy timberlands. Yet theft persisted afterwards. In 1881 the numerous trespassers on Federal lands were warned to cease their practices or face prosecution. In that year Charles Wuesher, a Federal agent, informed the Moss Point millmen that the laws were being violated in Jackson, Perry, and Greene counties. According to the marshal, large rafts of timber from public lands had been or were about to be floated to the coast mills. He warned the lumbermen that he would seize the lumber if the stolen logs were sawed by the mills. The Federal agent prevailed upon the millmen to require sworn affidavits from the logmen stating that no timber brought down the river had come from government lands. Millmen gave bond to the Federal authorities and agreed to withhold twenty-five cents on each log bought until the origin of the timber was determined.

In 1885 depredations were so numerous in southern Mississippi that several Federal agents were dispatched to the coast country to deal with the problem. One of these special agents, R. A. Vancleave, induced Fernando Gautier and Sons, L. N. Dantzler, and W. Denny and Company to sign an agreement to accept no timber cut from government lands. Vancleave reported to his superiors that public sentiment favored a policy of strict enforcement of Federal laws at that time. Despite his optimism, trespassers removed timber from hundreds of acres in Jackson County during the period 1884-1886.

In the late eighties there were signs that wholesale thefts might be gradually diminishing. Many of the lumbermen had acquired ownership of great tracts of timberland and thus became opposed to timber thieves. Nevertheless, after the Federal agents were removed from south Mississippi in 1892, depredations again became numerous in every locality where land remained in possession of the Government. Trespassing evidently continued to some extent until the last acre owned by the Government was transferred to private ownership.

The theft of timber was by no means confined to Federal lands during the late nineteenth century. To reduce trespassing, the Mobile and Ohio Railroad leased all of its Mississippi timberlands to A. C. Danner. From lands set aside for the Gulf and Ship Island Railroad large amounts of timber were stolen. Lands owned by northern syndicates were also plagued by theft. Indeed, in northeastern Jackson County logmen took timber from a Michigan landowner without ever being molested.[9]

The hopes of southern congressmen that timberlands would pass quickly into private hands through private entry were not immediately realized. In 1876, the year the act restoring entry was passed, the United States Government still owned approximately four million acres in Mississippi. Only 21,235 of these acres were disposed of in the fiscal year 1879. One of the principal reasons why sales were few was that an abundance of state timberland was obtainable for a fraction of the minimum price of Federal lands. Another reason was that land speculators from the Northeastern and Lake states had not as yet begun to invest in southern timberlands. In 1880, however, the editor of the *Northwestern Lumberman* received numerous inquiries concerning opportunities for lumbering in the South. In reply, the editor wrote that large tracts of southern timberland, ideally situated as regards both water and rail transportation, could be purchased at a cost of from $1 to $5 an acre. Other timberlands far removed as yet from transportation facilities were available at prices ranging from twenty-five cents to $2.50.

In 1880 the volume of sales of Federal lands increased, and during the years 1881-1883, 1,007,010.46 acres of land in Mississippi were acquired by individuals under the various land acts. As a result, Federal holdings in the state fell to 2,201,876.54 acres. Naturally, the first of these lands to be bought were those located near rafting streams and railroads.

Because of the rapid sales of land, the United States Land Commissioner began to advocate ending private entry, and he recommended that the lands be sold at their real value. Because of the swift disappearance of virgin forests, the commissioner by 1884 had come to believe that only moderate amounts of Federal timberland should be sold from time to time; he also favored the establishment of national forest preserves. Observing that 619,000 acres located in the Pearl and Pascagoula districts had recently been bought up, largely by outside interests, the editor of the Pascagoula *Democrat Star* in 1886 decried the rapid rate at which timberlands were passing into the hands of northern and eastern capitalists. According to the editor, the Pascagoula millmen, who owned little land, had previously obtained most of their timber from the national domain, but with its passing they would soon be at the mercy of western corporations.

During 1883-1884, Michigan investors were the largest purchasers of Mississippi lands. In the next two-year period many Chicago capitalists awakened to the opportunities for profitable investment. By 1887 most of the timberlands in Mississippi contiguous to railroads had been purchased from the Government or from individuals, and, as a result, prices of privately owned lands had risen from

$1.25 to $15.00 per acre. In order to obtain large blocks of land, buyers were subsequently compelled to take up land far removed from both rafting streams and railroads.

In 1887 two large groups of purchasers representing a capital of $2,000,000 traveled throughout all the Southern states where public lands were located. In addition, special excursions for lumbermen were initiated by the railroads. By this time some farsighted lumbermen were predicting that the South was destined to become the major lumber-producing section in the nation. Because of this new interest in government lands in the Southern states, the Federal Government sold through private entry 128,284.56 acres in Mississippi for $1.25 per acre during the fiscal year ending June 30, 1887.[10]

These Mississippi timberlands were purchased by both individuals and groups. The firm of Edward A. and Edward F. Brackenridge, for example, was said to have selected and located nearly 700,000 acres of pineland for its customers. On the other hand, Delos A. Blodgett was the largest individual purchaser of lands in Mississippi. He had entered the logging business in Michigan in 1848. Investing the profits from his business in timberlands, he built the largest sawmill in Michigan in 1878. In 1882 Blodgett began to purchase both state and Federal lands in Mississippi, and by 1906 owned 721,000 acres of long-leaf pinelands, most of them in northern Jackson, Wayne, Greene, Perry, Marion, and Pearl River counties. The Blodgett family sold large tracts of timberland in the years 1908-1912 to lumbermen at prices of $30 and $40 an acre and more.

Probably James D. Lacey was the largest single speculator in southern lands. He purchased land from the Federal and state governments, and afterwards sold it to other investors and to lumbermen. James L. Gates, another large operator, at one time claimed to own 300,000 acres of southern pineland.[11]

The rapid transfer of large tracts of land to a small number of owners soon produced a fear of land monopoly among Mississippians. It was obvious that most of the land had been purchased by investors who had no intention of erecting mills for the manufacture of lumber. Moreover, much of the land involved, especially in the region west of the Pearl River and along the northern borders of the long-leaf pine country, could by the use of commercial fertilizer be made suitable for farming. These potential farmlands, however, probably would remain uncultivated for years if acquired by a few people whose only interest was timber. Sentiment in the state which had once strongly favored the sales of lands to speculators now turned against them. In fact, it was Senator Edward Cary Walthall of Mississippi who introduced a bill to end disposal of the national domain by private entry. The bill passed without debate in the House.[12]

In Mississippi the Federal Government disposed of 2,640,469.18 acres during the period between June 30, 1877, and June 30, 1890. Thus United States lands in the state were reduced to only 1,407,480 acres in 1890, mostly located in the timber counties of southern Mississippi. After 1890 approximately 90,000 acres were donated to Mississippi for schools and colleges, and additional grants of swamplands were made. The bulk of the remaining public domain eventually passed into private hands through homesteading.[13]

Although the intention of the Homestead Act of 1862 had been to bring into cultivation lands upon which settlers could make an independent living, no limitation in the law prevented nonagricultural lands from being taken up. The act applied uniformly to rich as well as poor soils. In regions of infertile soils, little of the public domain was converted into small farms; instead, homesteading contributed materially to the concentration of landholding.

Had the provisions of the act been rigidly enforced in the pine country, little of the public domain would have passed to individuals. In the first place, to meet the residence requirement of five years would have been impossible, since most of the land was too poor to provide the means of livelihood. Secondly, the settler was forbidden to remove timber from his land except that actually cultivated and needed for improvements until full title had been acquired.

In the years 1866-1876 a small number of homesteads in Harrison, Jackson, and Copiah counties became farms. According to I. C. Enochs, the freedmen moved to the pine ridges in Copiah County, cleared a few acres, and destroyed considerable tracts of virgin timber. Nevertheless, little of the pine country, even when land could not be bought from the Government, was taken up by homesteaders in the period before 1876.[14]

The homestead acts were thus ill-adapted to the realities of the pine country, and they commonly stimulated fraud rather than farming. Such continued to be the case after 1880. Lumber workers were paid a small sum by lumbermen and naval stores operators to enter the land, and then after acquiring full title, to transfer ownership to their employers. When the process was completed, the only remaining evidence of occupancy was an occasional one-room shack surrounded by a clearing of uncultivated land. The settler had disappeared, and the title to the one hundred and sixty acres of land was held by lumbermen or others engaged in forest industries.[15]

In one typical instance, a naval stores operator, George Leatherbury, boxed the timber on hundreds of acres of government-owned land. Then, after learning that he might be prosecuted for his illegal acts, he furnished money for his Negro employees to enter the land. Apparently such cases were common. Special agent E. N. Ebbs in 1890, for example, reported that there were thirty-one false entries in Harrison, Perry, Jones, Covington, Marion, Hancock, and Jackson counties; and he requested permission of his superiors to suspend twenty-eight additional entries in Harrison and Jackson counties. One individual in Perry County, acting in collusion with a Federal land agent, acquired thousands of acres in virgin timberlands through dummy entries made in the names even of his oxen and of deceased persons.

When the construction of the Gulf and Ship Island Railroad opened up extensive areas of virgin timber country, outsiders are said to have moved into this area for the sole purpose of homesteading forest land, acquiring in this fashion large blocks in northern Harrison County. Reminiscence has it that four spinster sisters from Michigan filed claims for adjoining homesteads. By building one house that extended partly inside the boundary of each quarter section, they met the requirement of the law, gained clear title to the land, sold the timber, and returned to Michigan.[16]

In retrospect it is clear that Federal land policies designed primarily for disposal of agricultural lands proved unworkable when applied to nonfarming regions. Although interested only in timber, lumbermen and naval stores operators were compelled to acquire large tracts of land in order to obtain raw materials. When timberlands could be acquired only through the Homestead Act, fraudulent means were used by lumbermen and others to supply their needs. In the long run the policies pursued by the Government in the disposal of timberlands were harmful to the best interests of the nation. Without requiring the slightest effort at conservation, these policies permitted the removal of timber from millions of acres of land for which the principal and best use will always be the growing of trees.

CHAPTER VII

STATE LAND POLICIES
IN THE PINELANDS

W HILE THE GREATER PART OF THE PINELANDS of southern Mississippi was eventually transferred to private ownership by the Federal Government, the State of Mississippi also played a major role in land disposal by selling the large acreages that came to it through various types of donations from the United States. In many localities in the pine country the state became the principal landholder. From the swampland grant—to be discussed presently in more detail—the state by 1898 had received about 3,331,636 acres.[1] In addition, another large donation of approximately 837,584 acres, known as sixteenth-section lands, came into the possession of the state. Of this amount about 308,400 acres were located in the pine country. Federal grants to colleges came to 256,000 acres prior to 1883, and a later donation increased the amount to approximately 350,000 acres. Indirect grants of about 1,393,930 acres to railroads further increased the lands controlled by the state in the pine woods.[2]

As the result of slow sales of Federal lands and pressure from the states, Congress had in 1850 passed legislation granting land to the states to encourage reclamation of overflowed areas.[3] The Swampland Act was one of the most important land measures ever enacted by Congress, for through it a substantial portion of the national domain passed to the states. The act provided for transfer to the states of all lands that had been rendered unfit for cultivation because of overflows. Proceeds from sales of these swamplands were to be applied to reclaiming flooded lands by means of levees and drainage systems. When submitting its claims for such lands, a state could claim the whole of a legal subdivision of land if the greater part was wet and unfit for cultivation. By broadly interpreting the clause "wet and unfit for cultivation" Mississippi acquired thousands of acres of nonoverflow, dry, hilly lands.[4]

If the provisions of the Swampland Act had been strictly adhered to, nothing but the unsold lands in the Mississippi Delta could have been rightfully claimed by the state. In south Mississippi there were large tracts of actual overflowed land still owned by the Federal Government only on the Pearl, Homochitto, and Pascagoula rivers and their tributaries. Moreover, the construction of levees would do little or nothing to render most of these lands fit for farming. They were generally low, wet meadows, or sandy bottom lands, and largely unsuited for agriculture.

Given the choice of compiling a list of swamplands from the field notes of government surveys or of appointing agents to make its selections, the state adopted the plan that would enable it to obtain the maximum donation. The advantage of using its own agents was that large acreages might be acquired which had not been listed as swamplands in government surveys. The law authorized the

appointment of agents who were to receive $10 for each
section patented to the state and the same rate for fractions
of sections,[5] for lands not previously listed in government
surveys as swampland. As the agents' compensation
depended upon the quantity of land they claimed for the
state, section after section of pinelands was culled from
the national domain. In northern Harrison and in Marion,
Hancock, Perry, Jackson, Greene, and Covington counties
much of the superior timberland was claimed by the state
in this fashion. Hilaire Krebs, one of the state's land
agents, selected thousands of acres of forest lands in
Jackson County. He judged as swampland all lands over
which a boat could pass. It was said of Krebs that he drove
a work animal hitched to a canoe across thousands of acres
of high, dry pinelands, which he later claimed for the
state. His locations were disallowed by the Federal Government, but not until the state had sold large tracts. This
brought about a lengthy litigation which was not resolved
until the passage of the McLaurin Act in 1905 which confirmed titles to the purchasers of the Krebs lands.[6]

In 1852 the Mississippi state legislature enacted the basic
law for administration of the swamplands. This act provided that all swamplands lying east of the western base of
hills bordering on the Mississippi River from the Louisiana
line to a point opposite the mouth of the Yazoo River and
on the streams east of the western base of the hills contiguous to the Yazoo and Tallahatchie rivers, would be used
to improve navigation and to reclaim waste lands. These
lands were to be granted to the counties in which they were
located, with the proviso that the money derived from the
sale of these lands might not be used for improvement of
navigation unless Congress consented. In each county
there were to be appointed "a commissioner or commis-

sioners not to exceed three in number to be styled 'commissioners of swamplands'." These commissioners were empowered to sell land for not less than fifty cents an acre, to direct construction of levees and drainage canals, and to apply funds to improvement of navigation if Congress consented.

One of the early important legislative acts was the establishment of the Southern District of Pearl River, composed of the counties of Copiah, Hancock, Marion, Lawrence, and Simpson. Each of the boards of police of the several counties was empowered to appoint two commissioners to form a board to be styled the "Commissioners of the Southern District of Pearl River." This regional group was given power to fix land prices, sell lands, and allocate the funds derived from sales to "reclaiming and draining said swamp and overflowed [lands] by ditching, levying [sic] or removing obstacles from said [Pearl] river." Lands bordering the Pearl River and large acreages located on the tributaries of the Pascagoula and Wolf rivers twenty miles removed from the Pearl were given to the district. Lands on hills neither flooded by overflow nor poorly drained were included in the grant. The sale price varied from fifty cents an acre in 1852 to twenty-five cents, a minimum price after March 1, 1859. Thousands of acres of virgin timberlands were sold to single purchasers.

Except for a few minor differences, the system of land disposal in other counties was the same as in the Pearl River district. In Harrison County the board of police was empowered to sell at the low rate of five cents an acre to any person living on the lands or having them in cultivation, or who might "have erected buildings upon the same, or machinery of a useful kind." Such persons could thus obtain a section of virgin timber for $32. A purchaser might not enter more than one section until sixty days after his previous entry; but this limitation could be circumvented by acquiring additional lands in the names of others. Another minor restriction which provided that any given entry must be for contiguous tracts was obviously intended to prevent the selection of scattered superior tracts.

The liberal terms of the state swampland legislation as applied to Harrison County were responsible for the disposal of thousands of acres in the years 1858-1860. Millmen and logmen bought section after section near rafting streams. John Toulme and others, for example, acquired more than 30,000 acres on the Wolf and Biloxi rivers. Malcolm, Colin, and James McRae were heavy purchasers of land near Red and Black creeks. Donald McBean obtained many acres in both Harrison and Jackson counties. Indeed, so keen was the competition and so great the demand that most of the state-owned swamplands in Harrison County were in private possession by 1865.

During the decade after 1850 the state selected around 500,000 acres in Jackson County. An act of 1856 provided that the swamplands there be offered for sale at public auction in sections, half-sections, or smaller fractions at a minimum price of fifty cents an acre. The interest received from investment of the proceeds of these sales could be used for "improvement of roads and bridges." But this was made subject to the proviso in the first section of the act of March 16, 1852. Land sales were slow until 1858. In that year the minimum price was reduced by one-half. A scramble for timberlands resulted. John R. Dickin, Thomas Galloway, Garland Goode, and others were heavy buyers. By 1865 all the state's swamplands in Jackson

County selected prior to 1860 with the exception of those located in inaccessible areas were sold.

By 1859 land prices had been reduced to twenty-five cents an acre in some of the pine counties. This reduction was probably due to the slow land sales during the period when prices were fifty cents an acre, and to the Graduation Act of 1858 which decreased the price of unsold Federal lands that had been on the market for ten years or more.

The land policies of the postwar government of Mississippi were determined by economic conditions in the state as well as the land policies of the Federal Government. By acts of the state legislature in 1865 and 1866, minimum land prices in the Pearl River district were lowered to 12½ cents an acre, and in Perry and Jackson counties to ten cents. County boards of police were authorized to borrow any or all of the "Swamp and Overflowed Land Fund" on hand or to be collected and apply it to the relief of destitute families of soldiers. Later in 1868 all lands located outside of the Pearl River district were appropriated for the use of common schools. Harrison County land prices remained unchanged from their prewar rates.

The Pearl River Improvement and Navigation Company incorporated in 1871 was the successor of the antebellum Southern District of the Pearl River. All the swampland belonging to the state on the Pearl River and some areas on the Wolf River were donated to the company. In return for this grant the company was to make the stream and its tributaries suitable for steamboat navigation from Lake Borgne to Jackson, at least, and as evidence of good faith, it was to file a bond of $50,000 with the secretary of state.

Lands held by the Pearl River Improvement and Navigation Company were purchased by M. S. Baldwin, a real estate dealer who came to Mississippi from Chicago in 1872. The company had never earned the lands because it had failed to give satisfactory bond or to improve navigation as required by the state's grant. The state subsequently relieved Baldwin from the bond requirement on condition that he pay twenty-five cents an acre for the land. Baldwin claimed that he had paid the company president, R. A. Voss, twenty-five cents an acre, but that Voss had absconded with the funds, and the state thus received nothing. When Baldwin proceeded to sell the lands in large blocks to a few purchasers, the state claimed that titles to these lands were vested in the state since the company had failed to earn its donation. The result was much litigation over land titles. The state legislature tried to clear up the matter in 1890 by enacting a law providing that any citizen of the state holding land to which titles were derived from the Pearl River Improvement and Navigation Company could obtain a patent from the state, provided that the property had not been conveyed to other persons by the state, and that twenty-five cents an acre and all taxes assessed against the land were paid.

The rapid sales of swampland in large blocks produced a fear of land monopoly among the members of the state legislature. This led in 1877 to an act that limited to 240 acres the amount of swampland that an individual purchaser could acquire from the state. This limitation had no apparent effect on the rapid disposal of lands. In 1882 Governor John M. Stone of Mississippi stated that the state had issued patents for 225,000 acres in the last two years, and very little "of the swamplands patented prior to 1878" still remained. Although the state was believed to be entitled to about 387,00 acres of land, patents from the United States had been received for only 177,000 acres.

Land sales were fairly rapid in the late seventies and they boomed in the early eighties. Homesteading and Federal and state sales together transferred hundreds of thousands of acres of pineland into private hands, mainly those of speculators or investors. The state in 1884 owned no more than 47,154.51 acres.[7] Restricting individual purchases to 240 acres did not prevent the acquisition of large blocks of land by a single buyer. Quite generally, it would seem, millmen, logmen, and naval stores operators evaded the law by entering land under the names of other people, sometimes their employees, who for a small fee gave the actual buyers a quitclaim deed before the ink on the patent had dried. One logman of northern Harrison County acquired over 20,000 acres by this method. Another used the names of his oxen to secure state lands.[8] Marx Wineman claimed over 32,000 acres in Wayne County which had been entered under names of 135 residents of Shelby County, Tennessee. The entries, presumably instigated by him, had been transferred to him in violation of the spirit and intent of the law. Calvin Perkins in 1882, claiming to act under a power of attorney from a number of people, entered about 100,000 acres of land.

The most gargantuan evasion of the two hundred and forty acre law was carried out by Phillips, Marshall and Company. M. S. Baldwin, an agent of the company, employed James Hill, W. H. Gibbs, and others to procure the filing of entries under the names of a large number of persons who were paid small sums for signing applications and quitclaim deeds. The company thus obtained about 140,000 acres of pineland. According to testimony before a Mississippi legislative investigating committee, Swampland Commissioner J. M. Smylie and other state officials, including the governor, were aware of the fraudulent purchases of Baldwin and his associates. Smylie appears to have accepted extra fees for assisting Baldwin to acquire land illegally. The investigating committee recommended that suits be instituted against the company to recover the lands, but nothing was accomplished. In 1885 all the pinelands owned by the company were advertised for sale because of nonpayment of taxes. John Watson, a stockholder of the company, advanced the amount required for back taxes and became owner of the lands. Watson in 1886 sold them to the Delta Pine Land Corporation, successor of Phillips, Marshall and Company. The investigation of the latter company served to dramatize the rapid engrossment of state lands by a few people and brought legislation designed to prevent a recurrence of large-scale purchases.[9]

In the session of 1888 the state legislature passed the so-called homestead act. Under its terms any person who was head of a family or twenty-one years of age, a citizen of the United States, and had been a resident of the state for two years, might purchase not more than 160 acres of land for fifty cents an acre. Deeds of land purchased under the act at the instance of any individual or corporation with intent to evade the provisions were declared null and void. In 1890, however, the homestead act was repealed. Price of lands was fixed at $1.25 an acre, and the maximum purchase was limited to 240 acres "in one continuous body." Purchasers were required to be *bona fide* citizens of Mississippi; corporations and nonresidents were thus debarred from acquiring lands from the state. This type of legislation was an attempt to implement the Jeffersonian ideal of small independent, self-sufficient agriculturists; but as a large portion of the land was totally unfit for farming, the plan, however noble its objective, could only fail.

In 1892 the office of swampland commissioner was abolished, and the administration of all lands owned or controlled by the state except for Choctaw school lands was entrusted to the state land commissioner. The amount of land that might be purchased by an individual was now limited to 160 acres within a twelve-month period, and corporations and nonresident aliens were prohibited from acquiring lands. The commissioner in his first report asserted that the pinelands owned by the Federal Government were inferior in quality to those owned by the state, because state agents had secured the best lands. In his opinion laws designed to limit land holdings were without effect, for speculators continued to buy up "the best timberlands in the world" for $1.25 an acre while other states were selling land for $19 per acre, and timberlands were now held by syndicates that would not sell at any price. He reported that the state in the past had obtained little revenue from land sales. County boards of supervisors had been responsible for protecting state lands, but they had often failed to do so. As pointed out earlier, the spoliation of state timberlands was frequently carried on by the very people whose duty it was to protect them. The commissioner recommended that all lands owned by the state be surveyed, classified, and sold according to their true value. He also recommended methods to check depredations upon state timberlands.[10]

Only a few of the commissioner's recommendations were accepted. Agents were appointed to oversee the timberlands and suits were instituted against depredators. But in the southern counties indictments rarely resulted in conviction when tried before unsympathetic juries. Even when convictions were obtained, the light penalties exacted did not serve as deterrents. Trespass on Federal lands as well as on state lands continued until the last acre passed into private ownership.[11]

The state in 1896 claimed about 145,000 acres in the pine country, including 71,153.63 acres in the three coast counties. The commissioner admitted that he was unable to determine the exact amount owned by the state. Large acreages had been improperly placed on the tax roll before being sold, and were disposed of through tax sales.[12]

Besides the swampland donation which constituted the bulk of the land conveyed to the State of Mississippi by the United States, sixteenth-section lands and Federal grants to colleges accounted for about a half million additional acres. Under land laws of the United States and of Mississippi, the sixteenth section of each township of thirty-six sections, except in the Chickasaw Purchase, had been set aside for support of public schools. In many counties, however, the sixteenth sections had been permitted to pass into private ownership. In such cases land in other counties, often in south Mississippi, were sometimes given to these counties. As a result, such lands, known as "lieu lands," embraced a considerable area in the pine country. In Hancock County alone lieu lands amounted to 30,710 acres.

The disposal of sixteenth-section lands was, as in the case of swamplands, often characterized by fraud and mismanagement. The state in 1833 provided for the leasing of sixteenth sections for ninety-nine years.[13] Little if any of the land in question was sold outright, but the ninety-nine-year leases practically amounted to sale. In the pine country boards of supervisors, whose members were often employees of lumber companies, sometimes leased the sixteenth sections to their employers or political supporters for only a fraction of their true value.[14] The lumber companies, after thus acquiring sixteenth-section leases, proceeded to remove the timber. Eventually their right to cut timber on school lands was challenged by the attorney general of the state, who contended that a lessee could remove only the trees necessary to clear land for agricultural purposes. Suits brought against lumbermen finally resulted in a court decision stating that removing timber for purposes other than farming and personal use constituted waste and a diminution of the inheritance. After this decision the lease-holding lumber companies could not legally cut timber; but most of them appealed to the supervisors, who proceeded to transfer timber rights to the companies at small cost.[15]

During the 1890's four institutions of higher learning in Mississippi received grants of between 22,000 and 23,000 acres each from the Government of the United States. Three-fourths or more of this land lay in the timber counties of the southern part of the state. Most of the lands acquired by the University of Mississippi had formerly been part of a naval reserve tract set aside by Congress in 1858. The university retained its donations, but the other colleges sold their property to lumbermen between 1895 and 1902. Gregory Luce and others bought 25,000 acres in Jackson and Harrison counties. D. A. Blodgett and the Hemphill brothers were heavy purchasers in Perry County as were the companies of Eastman-Gardiner, J. J. Newman, Knapp and Stout, and the Sage Improvement Company. The average price received by the colleges was from $4 to $6 an acre.[16]

Lands obtained by railroads also passed into the hands of lumbermen. Slightly more than one million acres in Mississippi were donated to the Mobile and Ohio Railroad in 1850. About 478,619 acres of this grant were located in five longleaf pine counties, with more than 202,059 acres in Wayne County alone. For many years these lands found no sale at the high price of $2.50 an acre fixed by law. The railroad company retained large acreages in Wayne, Clarke, and Greene counties and sold the timber rights to lumbermen.[17]

In 1856 the Federal Government offered to donate to the Gulf and Ship Island Railroad 120 sections for each twenty miles of road constructed. The amount of land the company eventually received is uncertain; apparently, however, it was about 108,229 acres. This land the company proceeded to sell to such lumbermen and speculators as Delos Blodgett, J. J. Newman Lumber Company, Camp and Hinton Lumber Company, the firm headed by Pratt, Burton, and Hill, and to the land speculator J. H. Moores.[18]

Mississippi's policies of land disposal were shaped largely by economic conditions within the state and by the outlook of a people whose pursuits were agricultural. During the ante-bellum period the swamplands were recklessly disposed of, because pinelands were unsuited for an agricultural population and were thought to be worthless. In the postwar period the state sought to reserve its lands for farmers and to prevent the concentration of land ownership in the hands of a few. The efforts of the state to promote wide diffusion of land ownership in the pine country failed despite the laws that limited the number of acres that individuals could acquire legally. The agrarian ideas that inspired the Swampland Act of 1877 continued to dominate land legislation after 1900 and to some extent had an adverse effect on the economy of the state.

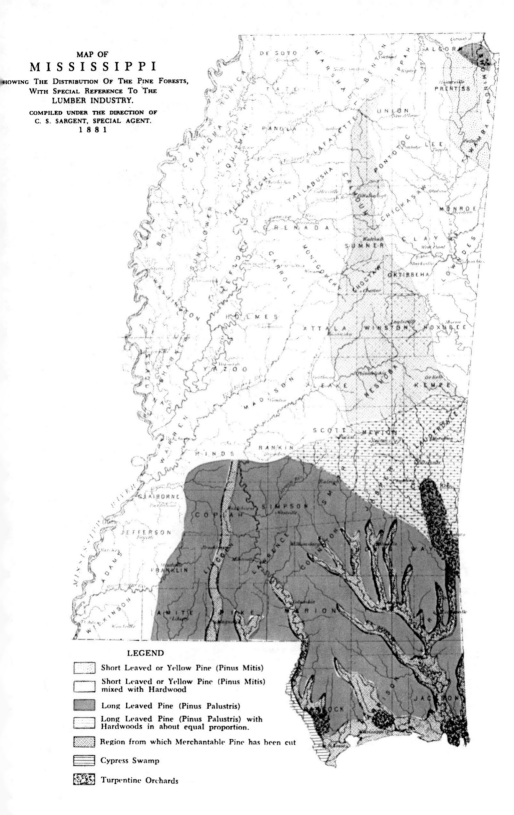

MAP OF
MISSISSIPPI
SHOWING THE DISTRIBUTION OF THE PINE FORESTS,
WITH SPECIAL REFERENCE TO THE
LUMBER INDUSTRY.

COMPILED UNDER THE DIRECTION OF
C. S. SARGENT, SPECIAL AGENT.
1 8 8 1

LEGEND

Short Leaved or Yellow Pine (Pinus Mitis)

Short Leaved or Yellow Pine (Pinus Mitis)
mixed with Hardwood

Long Leaved Pine (Pinus Palustris)

Long Leaved Pine (Pinus Palustris) with
Hardwoods in about equal proportion.

Region from which Merchantable Pine has been cut

Cypress Swamp

Turpentine Orchards

Map I

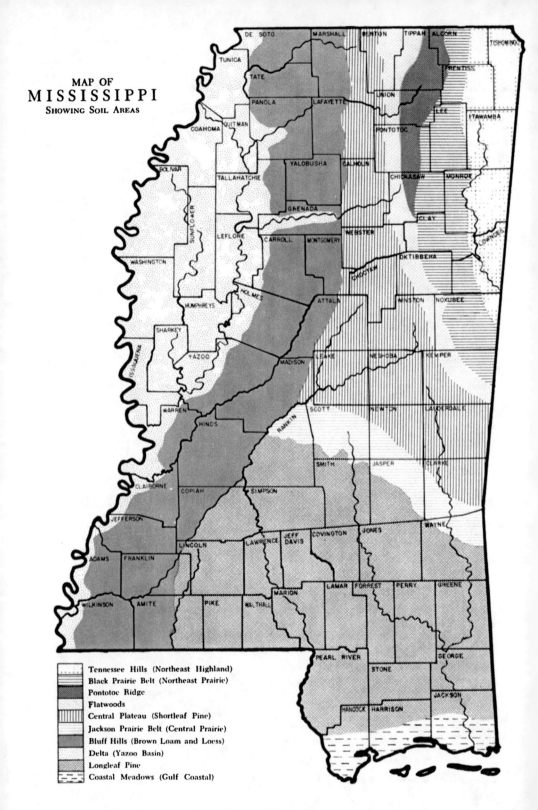

MAP OF
MISSISSIPPI
SHOWING SOIL AREAS

Tennessee Hills (Northeast Highland)
Black Prairie Belt (Northeast Prairie)
Pontotoc Ridge
Flatwoods
Central Plateau (Shortleaf Pine)
Jackson Prairie Belt (Central Prairie)
Bluff Hills (Brown Loam and Loess)
Delta (Yazoo Basin)
Longleaf Pine
Coastal Meadows (Gulf Coastal)

Map II

Figure 1   *Cole Pole Road Engine*

Figure 2   *Logging Train*

Figure 3   *Longleaf Sawmill*

Plate I

Crew of Sawyers Using Crosscut Saw

Plate II

Longleaf Pine Trees Tapped for Their Gum

Plate III

Fig. 1 *Using the Hack*   Fig. 2 *Cupping the Gum*

Figure 3 *The Still*

Plate IV

J. A. Simpson and Crew Beside Distillery

Plate V

Main Line Logging Engine and Loaded Cars

Plate VI

Lindsey Eight-Wheel Wagon

Plate VII

Clyde Two-Line Skidder at Work

Plate VIII

Clean Cutting of Trees

Plate IX

Westonia: Logging Village of the H. Weston Lumber Company, 1910

Plate X

CHAPTER VIII

LOGGING AND RAFTING
1840-1910

BETWEEN 1840 AND 1910, logging and rafting underwent few changes. Logging consisted of two basic operations; namely, felling the trees and hauling the logs to the banks of rafting streams. Until the late eighties axmen, commonly known as "choppers," felled the trees. Their basic tool was a pole ax having a single cutting edge and weighing about five pounds. In the seventies the double-bit ax with two cutting edges supplanted the pole ax. As long as axes were used, trees were cut about waist high from the ground, leaving a three- or four-foot stump. After notching the tree on the side where he wished it to fall, the axman did most of the chopping on the opposite side. At length, when only a small layer of wood separated the notches, the tree came crashing to the ground. It was then cut to the desired length.[1]

Expert choppers possessed great endurance and a high degree of skill. Many of them could cut the ends of logs with an ax about as smoothly as if they had used a crosscut saw. Axmen who could fell fifteen to twenty trees daily earned good wages. A few of the choppers employed by Calvin Taylor in the late fifties, for example, made more than $1 per day. The average axman, however, considered cutting ten logs to be a good day's work.[2]

Trees suitable for spars probably were the first to be culled from the longleaf and slash pine forests. As early as 1850, if not before, men moved through the back country in search of the long, slightly tapering trees so much sought after by navies of European countries. A special type of tree was required to meet rigid naval standards. The tree had to measure twenty-six to thirty inches at the large end, and eighteen to twenty-one inches at the smaller, with a length from seventy-five to one hundred feet. The amount of timber that could meet these exacting tests was limited.

The process of hewing spar timber into an octagonal shape required special skills of the axman. Some spars had eight sides for part of their length and four for the remainder. After the tree had been cut down, lines on the timber were marked off, and the chopper with a wide bladed ax, known as the broad ax, proceeded to hew away the excess wood. It was said of the expert choppers that they could hew a surface as smooth as if done by a planing machine. Only an expert could work up as much as one stick of spar timber within a day.[3]

"Square timbers" were logs of four equal sides hewn from the largest and finest trees in the forest. This timber was marketed in foreign countries, where it was used for heavy construction or sawed by mills into smaller dimensions. "Sawn timber" differed from square timber only in that it was sawed instead of hewed into shape. Square timber, while less valuable than spars, brought higher prices than saw logs. In the early eighties, sawn and square timber sold for eight to thirteen cents a cubic foot; after 1900 the price never fell below seventeen cents.[4]

Scattered throughout the longleaf forests were large quantities of standing deadwood. All the sapwood in these logs had been withered away by time, weather, and fire, and the remaining wood was all clear heart, often of heavy resinous content. Dead timber of this kind was excellent material for structures such as bridges requiring great strength and durability. The piney woods people, who called dead timber "light'ood," used it for sills, house blocks, fence posts, and rails, and sawed much of it into lumber. Its most important use came with the construction of railroads. Then a large number of people were employed hewing hundreds of thousands of crossties from the big light'ood logs.[5]

In the late eighties the crosscut saw replaced the ax as the basic tool for felling trees. (See Plate II.) Its coming was a major event, and people often gathered for miles around to witness a demonstration of what the new tool could accomplish. The first saws were unsuccessful, because the kerf clogged and choked the saw, and at the same time the resin of the yellow pine logs formed a thick layer of gum on the saw teeth that prevented them from cutting. Subsequent improvements, such as the inclusion of cutters, removed the sawdust, and the heavy film of gum was eliminated by the sprinkling of kerosene on the teeth and sides of the saw.

With the advent of crosscut saws came contract sawing at a stipulated rate per thousand board feet, the rate depending upon the size of timber, the thickness of stands, and the current wage scale. Adoption of the saw also led to the formation of so-called saw crews in which the foreman, a saw filer, supervised the operation. With the saw, trees were butted off close to the ground to obtain a fuller utilization of timber than felling with the ax had permitted.[6]

Bringing logs from the forest to the banks of streams before 1900 was accomplished by caralogs pulled usually by four yokes of oxen. The date when the first caralog appeared in Mississippi forests is a matter of conjecture. In the fifties or earlier, however, they were being used by Biloxi River and Bayou Bernard logmen. In 1852, for instance, B. L. C. Wailes observed that logs were hauled to streams with large cart wheels pulled by oxen. The original caralog wheels, according to Etienne Maxson, were of small diameter and had a tread about four inches wide. Modifications introduced somewhat later by Usan Vaughan, a slave of Nezan Favre of Pearlington, Mississippi, made the wagon practical for hauling logs in Mississippi. Vaughan's changes consisted of widening the tread and increasing the diameter of the wheels to more than seven feet. The wider treads helped prevent the cart from bogging down in low wet land, and the high wheel made possible the transportation of logs of any size.[7]

After 1900 the caralog was outmoded by the eight-wheel log wagon developed by John Lindsey of Laurel, Mississippi. Lindsey's wagon retained the wide treads of the caralog, but decreased the diameter of the wheels by more than one-half and increased the number from two to eight. Bolsters were built above the wheels to hold the logs on the wagon. The logs were conveyed to the wagon bed on skids, one end of which rested upon the ground and the other on top of the wagon. Oxen furnished the power to pull the logs from the ground to the wagon bed. Lindsey's eight-wheel wagon more than doubled the carrying capacity of the older caralogs.

For hauling logs oxen proved superior to mules and

horses with both the caralog and the eight-wheel wagon. In low wet country and boggy swamps they were less susceptible to bogging than were other draft animals. They also required less feed and, when not actually working, could be turned loose on the open range to graze without cost to their owners. Most of the oxen used in the piney woods were obtained locally from cattlemen when the steers were between three and four years old.

Ox driving was one of the oldest occupations in the back country. In this region it was not unusual to see a youngster twelve to fourteen years of age driving oxen hitched to a caralog before he possessed the strength to lift the log to the axle with the windlass. Many drivers seemed to have a special knack of getting oxen to obey. By talking to their charges and on occasion plying the whip, they obtained maximum performance. Others even with most brutal treatment and effort were never successful in getting good results from their teams. Patience was perhaps the outstanding quality of the successful driver.

The basic tools of the ox driver were the ox whip and the cant hook, and, later, the peavey. The whip, made of plaited cowhide, and eight to twelve feet long, was mounted on a slender stick from six to eight feet in length called the whipstock. The "popper," made of dressed deerskin, was attached to the free end of the whip and provided cutting quality. With the long whip the driver could strike a recalcitrant ox, regardless of his position in the team. But one writer has stated that

> The skilled use of the ox whip was not so much in its application to the beasts of burden to make them pull, though this was as much of an art as the use of the whip in driving loose herds, but was in its manipulation by the driver for his own entertainment and the enjoyment of whoever might be in hearing distance.[8]

An expert could bring forth from the ox whip sounds identical with the crack of a rifle or the boom of a cannon.

The cant hook, about five feet in length, was a wooden stock with a hook fastened to one end. Its principal function was to turn and manipulate logs. In 1858 Joseph Peavey, a Maine blacksmith, invented the peavey by combining the spike pole and the cant hook. The peavey was basic to both logging and rafting. The spike pole with a length of about sixteen feet consisted of a heavy hardwood stock with a sharp pointed steel spike attached to one end. It was designed to manipulate logs into position and to raise sunken logs to the surface.

Ox drivers, unlike choppers and sawyers, were paid a daily wage. In the ante-bellum period some of the drivers got $1.25 per day. For most of the years 1840-1900 average wages were fairly constant. After 1900 the wage scale gradually increased until the average by 1914 was around $2 a day. Pay varied from locality to locality, depending largely upon what an employer was willing to offer.[9]

Large log camps composed of fifty workers or more were virtually unknown before the construction of tramroads; an average logging operation for a locality required only a few choppers and drivers. The temporary camps were built near streams where water was easily obtainable for men and livestock. The huts, consisting of notched pine poles placed upon one another to a height of five feet and roofed over with shingles cut by the workers, had neither doors nor windows. The unsealed spaces between the poles gave little protection against the chill of winter's wind. The huts offered no more than a place for men to sleep and protection against rain and sleet.

After the logs had been hauled to the banks of streams they were given a distinctive brand by the owner. One logman might use a large number of different brands to indicate on what stream the logs were rafted or from whom his purchases of timber had been made. Ordinarily both logging and rafting were conducted by the same individuals, but not always. Many contracts covered only one phase of the process of bringing timber from the forests to the mills.

Sending timber to tidewater was common on most streams in the Gulf Coast country. Rafting logs was possible on any stream having banks four or five feet high where logs could be maneuvered. In the coastal plain country, where the average annual rainfall was from forty to sixty inches, the streams were frequently in flood, and logs in large quantities were moved to the mill in every season of the year. Many lesser streams were temporarily adequate for floating timber during seasons of heavy rain; in almost every locality of the pine country there was at least one small stream where logs could be rafted.

The date when timber was first sent to the coast cannot be fixed with certainty. By 1840 saw logs in small amounts definitely were being sent down the larger streams. The expansion of lumbering in the late forties must undoubtedly have increased the gathering of raw materials in the back country. The consumption of logs in 1849-1850 by Harrison County mills alone would have required the labor of a sizeable group of loggers and rafters.

By the late fifties the log business was of some consequence in the back country. Papers of the Taylor-Myers partnership reveal that many persons were cutting, hauling, and floating timber for them. At the same time a considerable number of logmen were selling timber to S. S. Henry. Hilgard, on visiting the coast country in 1860, discovered that fine lumber was being rafted down the Wolf River and Bayou Delisle to Huddleston's and other mills. He described the occupations prevalent in the region north of Gainesville as cattle raising, timber rafting, and tar burning.

Rafting logs to tidewater mills appears to have commenced again soon after the end of the Civil War. Rafting methods were generally determined by the size of streams and the swiftness of currents. The spike pole, the peavey, and the jam spike were the basic tools of the rafters. The jam spike, known in Mississippi as the jam pole, consisted of a heavy handle about five feet in length with a sharp spike inserted in one end. It was used primarily to pry loose logs that were jammed. On small streams and rivers, logs drifted down to the booms that were stretched across the mouths to catch them. In the Bayou Bernard area, loose logs came down the Biloxi rivers to the booms where they were sorted, made up into bull pens, and afterwards towed to the mills. This method of rafting was common on the Escatawpa, Jordan, and Wolf rivers, as well as on other small streams in the pine country. On the Pearl and Pascagoula rivers, and possibly on the Leaf and Chickasawhay, all streams with strong currents, other methods of floating timber were evolved.

When the freshets came, hundreds of men dropped their everyday tasks, hastened to the streams, and started the timber moving. As the logs started rapidly downstream, their bumping together could be heard for miles. Many obstructions that hindered the movement of logs had to be removed. High water often caused stream banks to cave in, throwing large trees across the water course and forming a barrier to moving timber. If the obstruction was not removed immediately, a jam formed above it with the resulting delay of the drive that could be costly if the high waters receded quickly. Logs stranded in eddies formed by sharp curves in the streams might also start jams if not pushed back out into the main current.

The task of keeping the drift moving was relatively easy when currents were swift and contained within banks. Where the banks were low, the high water tended to short circuit the curves and bends, carrying timber far out into the swamp, and such timber, if not worked back into the main channel before the stream returned to its normal stage, might become a total loss to the owner. Men riding logs or in boats used their long spike poles to maneuver the logs back to the main channel. But it was often impossible during a short period of high water to gather all the timber that had been stranded in the swamps. If the logs left marooned were near the stream banks, crews with peaveys picked them up and carried them to the stream. Ox teams were used to snake back to the channel those logs that had been washed far away into swamps and cutoffs.

Serious difficulties were encountered when low water found the drive a long distance from its destination. As the water level fell, jams became more numerous and the slowing current hampered and delayed the movement of the drive. If high water subsided quickly, most of the logs were apt to become stranded on sand bars and in shallow water to remain until further rains brought swift water. When low water threatened the drive, men worked night and day to get the timber to its destination.

Running logs in very small streams was difficult. In them the water level was likely to rise suddenly and subside quickly; timber must be moved quickly if it were to be moved at all. Sharp bends and curves common to small creeks caused numerous jams. Rafters equipped with long spike poles and peaveys walked along the banks and worked to keep the timber moving.

A log raft or crib accompanied the loose timber on its trip to the mills. On it the cook, usually one of the highest paid of the crew, prepared the food in iron pots over an open fire built upon its floor. He also took care of the bed-rolls and personal belongings of those who were engaged in rafting. Ordinarily, unless faced by low water, the crib was tied to the bank at night.

On the larger streams, methods more elaborate than the mere running of loose logs had to be devised. Before the construction of the large boom at Moss Point, logs destined there were assembled into large rafts above the mouth of Red and Black creeks on the Pascagoula River. A raft was built by uniting a number of cribs each containing from ten to fifteen logs. The logs in the crib lay parallel to one another and were held in place by a binder usually made of a hardwood pole. The cribs were tied together by ropes, holes for which were bored in both ends of the out-side logs of each crib. The ropes, about three feet long, were inserted in the holes and held in place by wooden pegs. The small vacant space thus left between the cribs lent flexibility to the raft and helped to prevent it from breaking up on striking an obstruction.[10]

A system of controlling logs less complicated and expen-sive than rafts was the "bull pen," which probably got its name from the pens used by cattlemen on the open range for marketing and branding. This method of sending logs to the mills was practical only in the relatively slow-mov-ing streams of tidewater areas. One of the main penning points was at the lower end of Dead Lake, where a boom made of large square timbers chained together stretched from shore to shore. In the center of the boom was a slot about fourteen feet in length that could be opened and closed at will. Above the slot was a runway upon which two crewmen stood. As logs passed through the slot underneath the catwalk, these men revolved them with their feet until the brands could be read. They then called out the brands, and other rafters, standing on chutes, guided the logs into the appropriate pens. The pens were constructed by tying the ends of several logs together to form an enclosure that would hold about one hundred logs.[11]

In general, floating rafts and bull pens were satisfactory on rivers having a rise of five or six feet. Such streams remained within their banks. Their currents were strong enough to move timber quickly downstream, and their depth of water was great enough to carry the rafts and bull pens safely over snags in the river bottom. A crewman rode each raft, steering it with a long pole having a flat board attached to the end. Often, however, he was unable to control the raft because of swift current. Moving at a rapid pace, out of control, the raft then would sometimes break up into sections if it encountered hard obstructions.

The length of time it took for rafts to complete their journey from the junction of the Leaf and Chickasawhay to Pascagoula depended on the swiftness of the current. If the water ran slowly, weeks were required to bring timber a distance of two hundred water miles or more from the headwaters of the Leaf, Bouie, and Chickasawhay to the mouth of the Pascagoula River. Rafting from the mouth of Red and Black creeks to the mills at Moss Point usually required from three to five days, but might take much longer under less than optimum conditions.

Occasionally high water came with little warning. In 1874 the rivers rose to a height not previously seen by living man. All the booms located at the mouths of the Pearl and Pascagoula rivers were broken up by logs pushed along by the strong currents. The heavy timbers on their way out to sea battered down the railroad bridges that spanned the mouths of the rivers. When the freshet subsided, logs were scattered all over the Mississippi Sound and in the low marshes adjacent to the shore line.

Log jams, although frequent, seldom required more than a few hours to break. But a jam at the Pascagoula boom in 1900 was unusual and probably the largest ever formed on the river. The weight of hundreds of thousands of logs pushed down many of the pilings that had been driven to form the boom. The height of the jam at its base was from fifteen to twenty-five feet, and logs piled upon one another for seven or eight miles upriver. Rafters with saws, axes, peaveys, and jam spikes cut a passageway for boats through the mass of logs in about two weeks, but over forty days elapsed before all the stranded logs were removed.[12]

By the early eighties lumbering had become the main occupation of the back country people. Almost everyone either directly or indirectly had come to depend upon logging and rafting for necessities. Living on small farms which produced only enough feed for a horse and a few head of livestock, the inhabitants devoted most of their labor to timber work. Farming was something to engage in during a part of the summer when lumbering declined. Bostick Breland, writing of the early eighties, said:

> Saw logs were the chief commodity of value, and they were floated down the various streams to Moss Point and other places where they were sold for a song. Few realized any cash money out of the business, but were satisfied if they got plenty of pickled pork, cheap family flour, coffee, brogan shoes, and a jug of joy.[13]

When the rains came, rafting took precedence over all other tasks. Men put aside their jobs, hastened to the creeks, and started the timber moving toward the mills. Long dry spells were often periods of tragedy, for with no employment, many went hungry. In dry weather men gazed at the sky in search of rain clouds, and with their approach hearts were made glad.

The physical hardships and discomfort that went with rafting were accepted without complaint. The spirit of adventure that came with running logs was felt by old and young alike. Youths looked forward with eagerness to the time that they would be permitted to ride a raft of logs to Moss Point. To be a member of a rafting party and to participate in a log drive was to youngsters one of the steps on the road to maturity. Old men, after their rafting days were over, never tired of recounting the experiences that were so deeply engraved in their consciousness. Most of them said that of all the tasks that they had performed, rafting was the best loved. A few were heard to say, after the last logs had gone down the river, that if they knew of another country like that which had once existed on the Pascagoula, they would go there at once.

Rafting was seldom dangerous in the Mississippi pine country. Northerners writing of hazardous rafting in the Lake states have stated that the jams were broken by finding and removing the "key log," and that to remove it was a dangerous operation. The southern rafters broke many jams, but were unacquainted with the key log. Drowning was about the only danger that threatened the riverman, and tragedies of this kind were few. Diseases caused by exposure to extremes of heat, cold, and rain accounted for more casualties among lumbermen than those resulting from bodily injuries.

Running logs, although at its peak in late winter and early spring, went on throughout the year whenever water was available. In the dead of winter, with temperatures below freezing, logs were ridden by men usually ill-equipped to withstand the inclemency of the weather. But in spite of long periods of exposure to cold weather, few became seriously ill. Indeed, men suffered less in the winter than in the summer because sleeping at night during the hot weather was all but impossible. In the low ground near creeks and riverbanks millions of mosquitoes literally filled the air. Lack of protection against the mosquitoes made malaria prevalent among the rafters; fortunate indeed were the few who escaped chills and fever. Occasionally when a yellow fever epidemic struck the coast, infected logmen carried the deadly scourge into the interior.[14]

In addition to the physical hardships experienced by rafters their business relations with millmen were often unsatisfactory. The method of timber inspection was in many instances said to have been disadvantageous to upcountry rafters and timbermen. For most of the period 1840-1915 logs were measured at the mill by log inspectors who were employed by the millmen. Up to 1900, before timber had become valuable and the supply scarce, only logs without blemish were marketable. Timber had to be free of knots, with no more than one inch of sappy wood. Logs that failed to meet the minimum requirement by only a small fraction of an inch were either rejected or else accepted at a price lower than the standard.

In the eighties rejections were especially high, averaging in some instances 10 per cent of the total delivery. The logmen were particularly incensed against the millmen, as many of the rejected logs were sawn by the mills and thus amounted to a gift. Many logmen tried to circumvent the strict inspection by concealing defects in their timber. As hollow knots on logs were almost always an indication that the wood inside was either rotten or otherwise defective, dishonest loggers drove wooden pegs in the small hollow knotholes to prevent air from escaping to the outside. This escaping air came to the surface of the water and created tell-tale bubbles which were always an indication that the log was defective. The inspector had to keep a sharp lookout for bubbles and wooden pegs.[15]

The general dissatisfaction of logmen over rules of inspection led to the establishment of the Pascagoula Timber District by an act of the legislature in 1886. The law provided for the election of a timber inspector by the voters of Jones, Jackson, Greene, Perry, and Wayne counties. This official's duties included measurement and inspection of timber at the Pascagoula booms. But this move to establish an impartial inspector subject to popular control was apparently unsuccessful, for in 1888 an inspector was appointed by the Moss Point and Pascagoula millmen. The right to appeal directly to the millmen the decision of the inspector was accorded the logmen.[16]

The control of timber and prices exercised by the Pascagoula and Moss Point millmen was threatened in 1884 when a group of Michigan lumbermen applied for a charter to establish a booming and rafting company on the Pascagoula River. The Michigan people were turned down by the state legislature, probably on account of the opposition of the coast lumbermen, who stood to lose their monopoly if the outsiders succeeded in their project. At this time few of the Moss Point lumbermen were unable to supply their mills with timber from their own lands. To permit a booming company to gain control of rafting meant that the millmen would be dependent upon outside interests for their timber supplies.

By the early nineties the enormous amounts of timber

coming downriver forced the Moss Point-Pascagoula lumbermen to devise a more elaborate method of booming and sorting logs. In 1893 all the large millmen at the mouth of the Pascagoula and a few upcountry logmen were issued a charter of incorporation for the Pascagoula Boom Company. The boom extended up East River from Moss Point five miles. Pilings spaced a few feet apart were driven midway in the river for the total distance. East River was then divided into two parts, one left open for navigable boats, the other enclosed for storage of logs. Sorting works adjacent to the boom were constructed so that logs might be separated according to owners. Almost all the logs except a small percentage that went down West River were caught in the boom. The boom company charged a small raftage fee on each log handled.

Construction of the boom reflected a growing expansion of lumbering. Although the number of mills at the mouth of the Pascagoula was decreasing, production concentrated in fewer manufacturing units continued to expand. This shift to large-scale manfacturing gradually brought about changes in the business dealings between interior logmen and coast lumbermen.

There was no uniformity in log contracts between loggers and manufacturers. In the early days down to 1900, and to a less extent afterwards, the logmen obtained contracts for delivery of logs to the mills at a set price per thousand board feet log scale. Usually the upcountry lumbermen received from the millmen advances in cash with which to hire workers and pay other expenses connected with gathering the raw materials. Upon delivery of logs to the mills the net return above advances was paid to the logman. In many instances the logmen obtained the funds from the millmen to buy timberlands, who for this service took a mortgage on the property. Eventually many of the debtors, when they were unable to pay off their loans, lost their land to the creditors. It was in this fashion and through the purchase of large acreages that some of the coast lumbermen acquired large land holdings.[17]

Most of the early logmen, although relying upon the millman to furnish their supplies and buy their timber, were fundamentally independent operators. They cut the timber from their own land, or purchased it from others, or took trees from the public domain. It was a general practice for many to contract the logging to others and to buy timber delivered to the stream banks. In this respect the logman exercised the functions of a merchant. In many localities the entire working population looked to the timbermen for employment, and, in some cases, for the actual means of subsistence. In the summer season, when little timber was rafted and lumbering was at a minimum, the workers drew rations from their employer and later repaid their debt in labor.

The timberman with the largest number of logs on the creeks was usually designated the "creek runner." By common consent the smaller logmen deferred to his judgment and paid him a raftage fee for running their timber. West Fairley, a Negro, was one of the many runners on Black Creek in the years 1872-1900.[18] During his early years as a logman, Fairley sold logs to Emile De Smet, and after 1884, he dealt with Lorenzo N. Dantzler, another Moss Point millman. Fairley conducted every phase of gathering the raw materials, buying logs, and rafting his timber and that of others to Moss Point. Fairley bought land from the state with money furnished him by Dantzler. Either through bad luck, mismanagement, or both, Fairley eventually became heavily indebted to his creditor. The land went to pay his debts, but Fairley continued to run Black Creek as an employee of the millowner.

Fairley's changed status represented a development that was occurring throughout the pine country. Sooner or later many logmen fell into debt, and their land as their most valuable asset went to pay off their financial obligations. But some of the timbermen of superior business ability became large owners and acquired wealth. A few of the back country logmen made the transition to large-scale lumber manufacturing.

Although many independent logmen continued to exist as long as timber was floated on the streams, by 1905 lumbering on the rivers was rapidly becoming a company-owned operation. In almost every locality there were mill company agents buying timber, conducting logging operations, and supervising rafting. Logging, even though on lands owned by the company, continued to be contracted to small operators.

Another factor that helped bring about company control in lumber manufacture was the growing scarcity of prime timber near rafting streams. As long as timberlands lay near the creeks and rivers, the small operator with little capital could get into business. But hauling logs with oxen became uneconomical if the distance exceeded four miles. The tramroad met the problem of distance hauling, but at an expenditure of capital that put it beyond the reach of the average man. One of the early tramroads to transport logs to the Pascagoula River was built by the Farnsworth Lumber Company. It ran from a high bluff on the river at Benndale to a large tract of virgin forest a few miles to the west. The logs were floated downriver from Benndale to the mill located at East Pascagoula. Other

Moss Point-Pascagoula millmen built logroads to provide their mills with logs.[19]

With the building of tramroads both the quality and sizes of timber sawn by the mills declined. Before 1890 few logs less than sixteen inches in diameter at the small end were marketable. But after 1900 timber inferior in both size and quality was utilized. The development of a foreign market for kiln-dried saps, lumber sawn from shortleaf pine, helped to increase the demand for small, sappy logs.

Utilization of the small logs increased the number of sinkers known as deadheads, far above the 5 per cent that had ordinarily sunk to the bottom of the streams when only superior timber was floated. In the years 1890-1915, and even earlier, thousands of logs were lost through sinkage. Deadheads, if they lay wholly submerged in the streams, remained in a state of perfect preservation, but when exposed to air and the work of insects they rapidly deteriorated. In the years 1905-1930 deadheads raised from the bottom of the streams constituted an important timber supply for the coast mills.

The expansion of the L. N. Dantzler Lumber Company reflects the shift to company control of every phase of lumber manufacture. After 1886 production of the Dantzler mills at Moss Point in some years was more than forty million board feet of lumber. With a large output and a heavy investment in mill machinery the necessity of keeping the mill in continuous production demanded a dependable, efficient system of collecting raw materials, and made too hazardous the old method of obtaining timber by contract with independent logmen, even though that method was simpler and perhaps more economical than owning land and timber. In 1909 the L. N. Dantzler Lumber Company got three-fourths of its timber from either land syndicates or from its own lands. Of the 249,088 logs received in 1912, only 5,000 were secured by contract. Logmen in the next year furnished 12,000 logs in a total of 142,361.[20]

Logging and rafting declined after 1910, mainly because most of the prime timber had been cut near the streams. With the construction of big mills in the interior and tramroads, rafting as a means of transportation of timber to the mills declined still further. By 1930 only a small amount of timber was being rafted to the tidewater mills. In the years 1890-1910 when rafting was at its peak, hundreds of men found employment. The business of gathering and sending raw materials to the coast mills brought economic opportunities to the people of a large section of the state where ways of earning a livelihood had previously been limited. The timber business enabled a few people to acquire capital, making it possible for them to become large producers. In a very real sense the marketing of timber bordering the streams was the initial step in the full-scale exploitation of the forests of the longleaf pine country.

CHAPTER IX

THE NAVAL STORES INDUSTRY

P RODUCTION OF NAVAL STORES is one of the older industries in the South. During the era of wooden ships, tar and pitch derived from cone-bearing trees were used to seal cracks in the flooring of vessels and to insulate ropes against the deteriorating effects of moisture. Hence came the application of the term "naval stores" to several substances derived from pine trees. Over the years a wide variety of products used for many different purposes, including camphene, turpentine, rosin, pitch tar, and rosin oil, has been developed by the naval stores industry. In the United States the raw material of naval stores comes chiefly from longleaf and slash pines, which, when wounded, exude from a network of small openings known as resin ducts, a mixture of turpentine and resinous acids called crude gum.

Methods of producing naval stores changed little in the years between 1834 and 1915.[1] Collection and distillation of the crude gum were the two basic operations. To be successful a naval stores orchard had to have a sufficient supply of water to operate the still as well as transportation facilities to carry the product to market. When locating an orchard experts regarded it as desirable to work timber in contiguous blocks. (See Plate III.) Supervision of labor was thus rendered easier and the cost of production was decreased. The first step in the gathering of crude gum was to cut boxes in the base of each tree about ten inches above the ground. These boxes collected the gum that flowed as soon as the tree was wounded above the box. The area thus wounded, or chipped, was known as the face. (See Plate IV.) The size of boxes varied with the diameter of the trees, but most of them were from three to four inches in width, seven inches deep, and twelve to fourteen inches in length. Such boxes could contain three pints of crude gum. Boxes were usually cut in the turpentine producing trees between December 10 and April 15. The tool used for this purpose was an ax having a blade twelve inches long and four inches wide with a beveled edge. Before operations started the land was surveyed and marked off into crops, each containing 10,500 faces or boxes. A crop was subdivided into drifts, each consisting of approximately 2,100 faces. The purpose of dividing a crop into drifts was to enable the woods-rider to exercise close supervision over the working of timber.

Cutting deep cavities into the trees was a crude method of collecting gum and one that was injurious to the timber. Often as many as four boxes were cut into the larger trees thus sapping their strength and causing many of them to die. Weakened by these cavities the turpentined trees were the first to be blown down by strong winds and the first to be destroyed by woods fires. Furthermore, much of the boxed timber if not sawed immediately made a poor grade of lumber. Lumbermen were right not to permit their timber to be worked for naval stores as long as the old methods were employed.[2]

When the resin began to flow in early spring, chipping of the longleaf and slash pines commenced and continued until late fall. Chipping or scarification of the tree consisted in cutting gashes in the tree above the box. The face from which the resin flowed downward into the container was formed by removing wood from the tree with a tool known as a "hack." The hacks, of different sizes, had flat steel blades 2½ inches wide which were bent into U-shapes measuring an inch between the sides. The blade of a hack was attached to a wooden handle eighteen to twenty-two inches long and about two inches in diameter. Fastened to the opposite end of the handle was a three-pound pear-shaped iron weight. This weight at the end of the handle enabled the workman to cut a deep streak into the tree with a minimum of effort.[3]

When using a hack the chipper, standing directly in front of the box, with two or three quick strokes removed a streak of wood ½ to ¾ inch wide and ½ to 1½ inches deep. Immediately after a streak was made the flow of resin was rapid, but it almost ceased within a seven-day period. For this reason a new streak was made by the chipper once each week during a season which usually lasted thirty-two weeks. Trees worked for the first year were called virgins and in the second year yearlings. No special names were used to designate timber that was chipped after the first two years. On such older trees, the puller (a tool similar to the hack, though with a longer handle and without a weight) was used to chip or pull the high faces. Virgins were the most profitable trees, since they produced both a greater quantity and a superior quality of crude resin. The flow of crude resin was highest in July and

August. In autumn, hardened under cooler temperatures, resin ceased to flow; large layers of material were left which produced a high proportion of low-grade rosin. Most operators abandoned orchards after working them for four or five years, though in some instances trees continued to produce for ten years or more. The faces of timber worked for more than eight years were often ten feet in height and could be chipped only by men on horseback.[4]

In general the chipper, who was the aristocrat among turpentine workers, was assigned one crop of boxes to work each week. Young Negroes looked forward with eagerness to the time when they would be given a crop to chip, for this achievement was a mark of manhood. Many of the superior workmen could chip from 12,000 to 17,000 faces a week, and a few were said to have put on more than 20,000. The average Negro turpentine worker started his work week on Tuesday morning and completed his task by Friday noon or early Saturday morning. A Negro who finished early either helped another complete his assignment or remained idle in the quarters until the following Tuesday. James B. Averit, who was born and reared on an ante-bellum turpentine plantation in North Carolina, stated that chippers being employed in large wooded areas out of range of anything amounting to close supervision had to be stimulated to put forth their best efforts. Therefore, the slaves were assigned such a quota of trees to chip as would allow them a day or more of free time in each work week. Moreover, premiums were given for good performance. The custom of assigning a quota of trees that could be chipped within four or five days of each week became well established during the ante-bellum period and has continued down to the present day.

Because the rate of flow of the resin depended on the

frequency of cutting the streaks, the chipping process had to be supervised closely by woods-riders. Lazy chippers were apt to put streaks on all the trees in a given area except those in the center in the hope that the woods-rider would fail to detect their omissions.

The rates paid to chippers varied from time to time and place to place. In almost every instance rates were higher in Mississippi, Louisiana, and Texas than in the states farther east. No doubt the supply of labor accounted for the difference in the pay scale. In 1896 chipping rates in eastern naval stores states ranged from forty-five to fifty-five cents per thousand faces, and in Mississippi, fifty-five to sixty-five cents. After the turn of the century rates slowly increased, so that by 1909, eighty cents and more per thousand was common in Mississippi. It is said that in some instances unscrupulous operators lowered their workers' pay by making the crops oversize and then crediting the chipper with having worked no more than the standard crop of 10,500 boxes. This dishonest practice was known as "loading the boxes."

The turpentine boxes usually filled about once every three weeks, though cool weather or droughts might lengthen the time. Negro dippers (often called derisively "sapsuckers," "sorelegs," and "tarheels" by others of their own race) emptied the boxes about seven times each season. Employing a flat trowel-shaped blade about 7½ inches long and 5½ inches wide, the dippers scooped the gum from the box (or cup) and emptied it into a pail or keg. From the pail the gum was poured into scuttle-headed barrels, and these when filled were rolled on skids up into the wagon. Negroes possessing great physical strength and endurance might dip as many as six or seven barrels each day. Much of this work was done, however, by women and children, as most adult Negro men considered it a low-caste job. A large amount of gum that hardened before reaching the box formed a layer of encrusted material on the face. This material, known as scrape, contained about half as much turpentine as the "dip" and yielded rosin of a darker color. After the last dip in late autumn, the scrape was removed with a rectangular blade from four to five inches wide.

Most of the distilleries were located near transportation lines in close proximity to the orchards. Until about 1910 the stills were usually of small capacity. Known as twenty or twenty-five-barrel stills, they processed about thirteen barrels of crude gum at one charge. The distillation of a single charge took from 2½ to 3½ hours. As the first step in the making of turpentine, the crude gum was emptied into the trough leading to the kettle and connecting the cap to the condenser. After the still was charged, the fire was started; its intensity was regulated by the noise of the cooking gum. The gum, after being brought to a boiling point, melted rapidly, and vaporized. The vapor moved from the kettle to a tube called the "worm" which was surrounded by a tank of cool water. In the worm the vapor condensed to a mixture of turpentine and water, and from there flowed into a tank. The turpentine, lighter than water, was then siphoned off from the upper part of the tank.

Rosin, a by-product of distillation, was drained from the bottom of the still through sets of screens into a vat sunk into the ground. These screens filtered chips and other foreign matter out of the rosin. After remaining in the vat for a period of a few minutes to an hour, the rosin was drawn off into barrels, where it cooled into a solid state. The methods of distillation employed in south Mississippi remained about the same for more than seventy-five years.[5] (See Plate V.)

In the late colonial period North Carolina was the chief source of naval stores for England. Following the Revolutionary War, the market for American naval stores collapsed, and the industry entered a long period of depression which lasted for more than forty years. A revival took place in the late thirties following the invention of the copper still in 1834 and the development of new uses for the products. In the twenty-year period following 1840 the demand for naval stores expanded, because of the use of turpentine in the manufacture of India rubber goods and varnish and as an illuminant. Camphene, a refined product of spirits of turpentine, was dangerously inflammable but possessed an illuminating quality rivaled only by sperm candles.[6]

Before 1840 only small quantities of naval stores came from any state except North Carolina, the center of the industry. Production in south Mississippi, for example, was only 2,215 barrels in that year. The growing demand for naval stores, however, brought about an expansion of the industry in the South and Southwest during the decade 1840-1850. In 1842 Fairfax Washington, a former naval stores operator of North Carolina, tapped a number of trees near Mississippi City to determine whether Mississippi pines were suitable for naval stores. His experiment was successful. Each tree boxed in November, 1842, produced a quart of crude gum by the following January. Washington then proposed to erect a distillery and offered $2.50 for each barrel of crude gum delivered to his still. This would, he claimed, afford profitable employment of labor, as one worker could collect two hundred barrels of crude gum in a twelve-month period.[7]

Despite Washington's efforts, the industry expanded

slowly, though longleaf and slash pines grew in abundance near navigable streams. Few workers were acquainted with the necessary skills, and capital was apparently scarce. In 1846 the editor of the Gainesville *Advocate* stated that while lumbering employed a large capital and many workers, the greatest resources of the pine country, tar and pitch, were as yet undeveloped. He predicted that naval stores would one day become an important industry in the pine country of Mississippi.[8]

In the late forties a few distilleries were set up in the coastal area. Nathaniel Mitchell in 1847 or early 1848 built a fairly large still at Napoleon, a few miles above the mouth of the Pearl River. Francis Leech and I. B. Ives, a cotton planter from Carroll County, served as managers of this enterprise. During two weeks of June, 1849, 2,500 gallons of spirits of turpentine, 600 gallons of camphene, and 250 barrels of rosin were manufactured. Nevertheless the business was abandoned in 1849 either because the cotton hands employed were unfamiliar with working the pine timber, or else because of a duel between Ives and Mitchell.[9]

By 1850 naval stores were being produced on a small scale in the three Gulf Coast counties, Harrison, Hancock, and Jackson. In Hancock County, for example, the value of the output of four small distilleries was $15,280. In Jackson County, Tom Galloway worked extensive orchards up the Pascagoula River at Brewer's Bluff. Abram Galloway, one of the slaves who aided in construction of the plant, later related that the still had been located on a high bluff, and that it had taken ninety days to build a double-track road from the top to the landing on the river. Cars loaded with turpentine and rosin had been moved on the tracks from the bluff to the banks of the river by means of a hand windlass.[10]

Expansion of the naval stores industry in Mississippi following the Civil War was slow, partly because an adequate supply of capital, skilled labor, and transportation facilities were lacking. According to the United States Census of 1870 production in Mississippi was less in that year than in 1850. But with the construction of railroads and the broadening of markets after 1870 the industry began to grow. In the late seventies a few distilleries were erected along the Mobile and Ohio and the Louisville and Nashville railroads. In addition several operations were started on the Pearl, Pascagoula, and Biloxi rivers.[11]

While the availability of transportation obviously was a limiting factor in the expansion of naval stores production in Mississippi, the lack of laborers with the necessary skills to work timber was another handicap. There were few Negroes or supervisory personnel before the late seventies in Mississippi familiar with production of turpentine and rosin.

By 1880 a few experienced operators from Alabama had drifted to Mississippi and with a few natives of the state established orchards. On the Escatawpa, Biloxi, and Chickasawhay rivers, and between Napoleon and Gainesville on the Pearl, such orchards were worked in that decade. Others were established in the timber areas adjoining the Mobile and Ohio and the Louisville and Nashville railroads. Most of the establishments of this type were small: the combined value in 1880 of the production of all eleven distilleries was only $97,000. Yet the census statistics of 1880 revealed that the industry was moving south and west. In the 1860 census the Carolinas had produced 90 per cent of the country's naval stores, but only 62 per cent in 1880. This trend became even more pronounced in the next two decades.

In Mississippi a slow but steady growth of the small naval stores industry during the 1880's resulted from three basic causes: virgin timber was becoming scarce in the Carolinas and parts of Georgia; construction of railroads opened up areas of Mississippi formerly inaccessible; and the markets continued to expand at nearby New Orleans. At State Line, Mississippi, J. C. Orrell and the Averys were among those who manufactured and shipped considerable quantities of turpentine and rosin down the Pascagoula River to tidewater. Orrell had been one of the first operators in the business and had acquired large areas of timberland. In northeastern Jackson County on the west bank of the Pascagoula River, George Leatherbury was a large turpentine producer. By 1886 the naval stores industry had developed to such an extent along the Pascagoula that a regular line of small river boats was engaged in bringing the products to market.[12]

The naval stores business in the longleaf pine areas adjoining the Mobile and Ohio Railroad was probably more highly developed than elsewhere in Mississippi. Most if not all of the lands owned by the Mobile and Ohio Railroad were leased to naval stores men in the early eighties. De Soto, Shubuta, State Line, Waynesboro, and Buckatunna, all located near the railroad, became small production centers in the eighties. From these towns 39,870 barrels of rosin and 236,220 gallons of turpentine were shipped by rail in 1895. At that time rosin brought slightly more than $2 per barrel and turpentine between thirty and thirty-five cents a gallon. State Line, producing 97,272 gallons of turpentine and 16,212 barrels of rosin, was then perhaps the largest manufacturing center of naval stores in Mississippi.[13]

In 1890 the Carolinas and Georgia with 623 distilleries were still the leading naval stores producing states. Mississippi had only twenty-four distilleries, and the other Gulf states including Florida had no more than twenty-three among them. The completion of the New Orleans and Northeastern Railroad and the Gulf and Ship Island Railroad was a powerful stimulus to the growth of the naval stores industry in Mississippi. Annual production along the New Orleans and Northeastern in 1891 and 1892 was approximately 15,000 casks of turpentine and 75,000 barrels of rosin.[14]

In the late nineties migration of both capital and labor from the Eastern states to the longleaf pine belt was especially heavy. Along the railroads in Mississippi, on the average, one small distillery for every five miles was erected as a result. An acute shortage of labor for the booming industry developed and a large migration of Negroes from Georgia and the Carolinas to Mississippi occurred. Naval stores producers who had exhausted the timber supply in other regions transferred their operations to the Gulf states, bringing their tools and labor with them. Trainload after trainload of Negro families arrived in the virgin pine country of south Mississippi. Many firms which had originated in North Carolina moved first to Georgia and then to Mississippi. Some of them continued to move west, ending up finally in east Texas. There, having no other place to go, the naval stores frontier came to an end.

The migration of the Carr brothers, who were small producers, from Georgia to Mississippi reflected the general movement of naval stores westward in the years between 1895 and 1905. The Carrs, John, Henry, and Arthur, maintained a small business in Georgia and, when timber grew scarce, they prepared to move to the virgin timberlands of southern Mississippi. In the fall of 1896 one of the Carrs arrived at Bond, a small town on the Gulf and Ship Island Railroad, and purchased a three-year turpentine lease on approximately four sections of virgin timberland. In that area most of the timber leases were being acquired from homesteaders in 160-acre blocks or from the railroad company. In the spring of 1897 the Carrs arrived at Bond with their equipment and twenty-six Negro laborers. They worked ten or eleven crops and produced about forty-six barrels of turpentine for each crop. When cups were introduced the crop yield rose to between sixty-five and seventy barrels, with three barrels of rosin for every barrel of turpentine. When leases were no longer available near Bond, the Carrs moved to a new location.[15]

Among the newcomers from the East was M. J. Bethune who came directly from North Carolina to Bond, acquired timber leases, and worked about fifteen crops. Bethune remained at his first location for a few years until most of the timber had been exhausted. In the same vicinity were many other small operators, Kennedy and Company, the Campeonois firm, O'Neal and Flurry, and others. Brasswell and Bullock, small Georgia turpentine men, brought their labor and equipment to Nugent, Mississippi, in 1898. In 1900 W. B. Gillican, destined to become one of the largest naval stores operators, was in business near Bay St. Louis. The Standard Turpentine Company had large orchards in southern Harrison and Hancock counties. At Caesar, R. R. Perkins with fifty crops was probably the largest single producer of naval stores in Mississippi in 1905. On the Pascagoula in 1899, 50,000 barrels of rosin and 15,000 gallons of turpentine were shipped to market.[16]

In 1899 there were 145 naval stores plants with a capital investment of $798,000 in Mississippi, employing 2,633 workers. Five years later the number of plants had decreased to 124, but the number of workers employed in them and the value of their total product were greater than in 1899.[17] The decrease in the number of plants with an increase in production and invested capital reflects the fact that units of operation were becoming larger. Indeed both the plants and the number of crops worked per firm from the beginning had been larger in Mississippi than in Georgia. This difference in scale of operation may be accounted for in part by the concentration of ownership of large landholdings and the larger average size of timber in the Mississippi pine regions.

In the years 1895-1905 conditions in Mississippi were favorable for small operators having little capital to invest. The railroads were being constructed through a virgin pine country where homesteading had been relatively heavy. During these years small turpentine men with limited capital were able to acquire timber leases or purchase land with comparative ease. Many of the lumber companies not familiar with the damage to timber caused by the old methods of turpentining leased land to small operators. After observing that injury was done to the tree by turpentining and that the woods fires had devastating effects on worked timber, many lumbermen ceased to permit their timber to be leased as formerly. Since most of the timberland was owned either by investors or by lumbermen, the small independent operators who were not connected with the few large naval stores firms were forced to suspend operations after the timber belonging to the small landowners had been exhausted. In the years between 1900 and 1905 the small turpentine operators began gradually to disappear. Naval stores production became increasingly concentrated in the hands of fewer people.[18]

The damage done to the timber by turpentining was eventually lessened by substituting a metal turpentine cup in place of the box cut in the tree. W. W. Ashe is credited with having made the first systematic attempt, at Bladenboro, North Carolina, to devise a more efficient method of collecting gum. Ashe's efforts and the contributions of Dr. Charles Herty eventually made the cup suitable for general use. Besides decreasing the damage to timber by requiring only a surface wound, the cupping system added to the value of the crop by increasing the quality of rosin and the amount of turpentine produced. Between 1905 and 1910 these cups came into general use in Mississippi.

The first turpentine cups were made of clay, but a later and improved type was fabricated from zinc and tin. The cups were held in place against the tree by a nail on the underside, and on top by a thin strip of tin, known as the apron, which was inserted into the tree and sloped downward. This apron induced the gum to flow from the face of the timber into the container. After the timber had been worked for one season, the cups were raised about twenty-four inches to a position immediately above the old face. Placing the cup close to the streaks was advantageous as the volatile oil, having only a short distance to flow on the face, would not be lost through evaporation.[19]

Although many lumbermen were reluctant to permit their timber to be worked for naval stores, the slow, steady increase of rosin and turpentine prices stimulated production nevertheless. Eventually, it became profitable to work trees for naval stores three years before cutting operations began. Felling the timber immediately after it had been utilized for naval stores reduced to a minimum the damage caused by turpentining and in many instances brought to the lumberman a fairly large profit. A few of these land-owners were able to pay for their timberland through such auxiliary naval stores operations, thus in effect securing raw materials for their sawmills at little or no cost. In fact, a part of the immense timber holdings owned by the L. N. Dantzler Lumber Company was acquired through their naval stores operations.

As a rule the large lumbermen eventually operated their own naval stores business or leased their timber to large turpentine corporations. This development brought an end to the business of many of the small independent operators, who must, to continue as producers, own their own land or sublease timber from the large naval stores firms. The Finkbine Lumber Company, a northern concern, became one of the largest producers in Mississippi. This company employed J. A. Simpson, a Georgian, to assume full responsibility for its naval stores business. Starting with twenty-five crops, he increased production as mill capacity was enlarged. Under his direction, the timber was worked for naval stores for three years before felling. When the company erected a second sawmill at D'Lo, Mississippi, it became necessary for Simpson to increase the amount of naval stores timber to seventy crops. In 1914 the company was employing about three hundred Negroes in its turpentine orchards alone.

The Finkbine Lumber Company was only one of many Mississippi lumber firms that branched out into naval stores production. In 1903 the Bond Lumber Company, another northern firm, produced 2,500 barrels of turpentine from forty-one crops. The company worked its timber for only two years in order to obtain maximum production and to minimize the damage to trees. The Denkmann Lumber Company, with large land holdings in Mississippi and east Louisiana, also operated a big naval stores business. The majority of large millmen in south Mississippi ultimately established their own naval stores firms.[20]

The years between 1905 and 1909 saw a continuation of the trend toward concentration of the output of naval stores in the hands of the landowning lumber companies. In 1908 there were approximately ninety-four operators in Mississippi, but both output and number of laborers were less than in 1905. Moreover, the number of laborers was less than in former years. This reduction in the naval stores business reflected a slight decrease in the amount of prime timber available. But the most important cause for a further decline in 1909 can be attributed to lower prices, which discouraged the investment of capital in naval stores. At the beginning of the 1909 season the number of plants in operation dropped to sixty-four; of that total twenty-seven were owned by corporations. Concentration of ownership in the naval stores industry was exemplified by the fact that 4.7 per cent of the plants produced 34.5 per cent of the output.[21]

By 1910 naval stores were being produced on a small scale by new wood reduction plants using the fat light'ood stumps of longleaf and slash pines as raw materials. The first two successful commercial plants of this type were located at Gulfport, Mississippi, and Brunswick, Georgia. These two factories, known as yaryan plants after their inventor, H. J. Yaryan, produced poorer grades of rosin from light'ood of slash and longleaf pines than those plants utilizing the resin taken from growing pine trees. Nevertheless products of the wood reduction plants eventually became strong competitors of gum naval stores.[22]

The naval stores industry, because of its relatively small size and the restricted area in which its production was centered, was subject to artificial influences governing prices and marketing. At an early date Savannah, Georgia, became the naval stores center of the world. Here were located the large export houses and the factorage firms which moved the naval stores to all parts of the world. The Savannah Board of Trade, composed of the larger firms, met daily, issued market reports, and fixed prices. Because of the members' connection with markets throughout the world and their heavy financial interests in producing firms scattered all over the South Atlantic and Gulf states, and because of the absence of substantial competition in foreign countries, the price fixed by the Savannah Board of Trade became the prevailing market price.

The Mississippi naval stores industry declined between 1915 and 1926, despite the phenomenal price level. With few exceptions production was concentrated in the hands of several large landowning lumber companies. Only in the counties of Perry, Pearl River, Greene, and a few areas of Stone, Harrison, Simpson, and Rankin counties were naval stores operations in existence even on a modest scale. This striking decline of the gum naval stores industry was the result of the elimination of the virgin timber. After 1924, the combined output of Mississippi, Louisiana, and the Carolinas accounted for only 6.52 per cent of the national production.[23]

Naval stores output reached an all-time low in Mississippi during the depression of the early thirties. With the virgin timber gone, production of crude gum was restricted to small tracts often located miles apart, a condition which materially increased the cost of gathering raw materials. Because of this scarcity of raw materials, the scope of operations was considerably reduced. Indeed, a thirty-crop operation was regarded as extremely large in this period.

New methods of distillation introduced or devised in the late twenties eventually outmoded the old fifteen- to twenty-barrel still by concentrating the refining process in a few large units. At the same time the gathering of crude gum became widely diffused, generally among a large number of small operators. To some extent ownership of the distillation equipment and the harvesting of crude gum were separated after 1930. Large numbers of small land owners performed almost all the labor in the production of crude gum, which they sold to the operators of the large distilleries. By 1950 there were fewer than six large stills in Mississippi. Most of the crude gum produced in the southern part of the state was transported to large distilleries located in Mobile and elsewhere.

During the period 1850-1915 the technique and methods of production and marketing in the naval stores industry underwent few changes. Tools used to work the timber and the method of distillation with few exceptions remained the same for the entire time. The marketing of naval stores products and, to some extent, the production of crude gum were controlled by the factors. Moreover, the concentration of ownership was much more pronounced in naval stores than in the lumber industry. The main reasons for this consolidation were that the naval stores were produced in a comparatively smaller geographical area than was lumber, and the value of the total annual product was much less. Control was undoubtedly easier to establish and maintain in an industry undergoing few if any radical technological changes.

CHAPTER X

LIFE IN THE TURPENTINE WOODS

ALMOST ALL OF THE PHYSICAL LABOR involved in naval stores production was performed by Negroes. Racking boxes and chipping were jobs over which the Negro exercised a near monopoly. In time Negroes came to bear the same relation to naval stores as their ancestors did to cotton in the ante-bellum South. Over the years the turpentine Negroes evolved a distinct society, rarely communicating with others except those of their own specialty. Negroes employed in agriculture, sawmills, and other occupations were conscious of the distinction between themselves and men who worked the pine trees, and usually sought little contact with them.

Because of his low income, his utter dependence on the white man, and his slave-like status in relation to his employer, the turpentine Negro was considered inferior by others of his own race. Although the standard of living of Negroes engaged in agriculture was perhaps on the same level as that of those who worked in the naval stores industry, the farmer, whether he was a renter or a share-cropper, had more personal freedom.

The sawmill Negro, who often held a position of responsibility in the mill, ordinarily enjoyed a higher standard of living than the turpentine Negro. Besides, he lived in close proximity to whites and tended to accept the standards of his superiors. Moreover, the sawmill Negro lived near lines of communication and was not continuously isolated as was the turpentine worker, who inhabited villages often miles away from railroads and towns. His very isolation tended to develop in him social attitudes and ways which deviated from those of the rest of his race. The uniqueness of the turpentine workers was clearly recognized by the operators, who, when establishing camps, always located them away from the quarters of the sawmill workers and the Negroes engaged in other occupations.

To the turpentine worker, emancipation from slavery had not brought the fruits of freedom. He simply had exchanged his lot for a different system of economic bondage in which the social security provided by slavery had disappeared. Often subjected to a ruthless economic exploitation, he was assured of employment only so long as a profit could be gained from his physical exertions. When no longer able to chip the timber because of physical infirmities or old age, the turpentine worker was cast aside to shift for himself. Such unfortunates usually died of starvation or disease or became dependent on relatives or associates for subsistence. Where the care of the aged and diseased is concerned, slavery was perhaps more humane than the industrial system used in the Mississippi naval stores industry.

The turpentine Negro enjoyed better living conditions

under the big operators who, after 1906, employed the greater part of the workers engaged in the industry. Moreover, the lot of the Negro seems to have been easier in the Gulf states, where labor was scarcer and the scale of individual operations was larger than in the South Atlantic states. Yet a shortage of labor did not always produce humane treatment. Often when the demand for labor was great, naval stores operators prevented the movement of Negroes from one location to another by force. In such cases guards were stationed on the few roads leading away from the Negro quarters to dissuade workers from leaving.

The average operator in the turpentine industry had two basic concerns in dealing with his employees—namely, that he have an adequate number of workers and that he obtain a profit from the worker's labor. As the turpentine business was not profitable to the average operator, the two basic considerations often resulted in his abrogation of the fundamental rights guaranteed the Negro by the Government. In some instances Negroes were subjected to coercion, brutal treatment, and other abuses.

Competition for labor among the naval stores men was particularly keen in the years between 1895 and 1906, when the industry was expanding rapidly in the Gulf states. Men often stooped to any means in order to obtain a labor supply. There existed a general agreement among the turpentine man (which few of them kept) not to recruit labor from one another; to be caught violating the agreement was a dangerous business. Many of the operators used trusted Negro employees to go into their neighbor's camp in search of workers. If his recruiter was caught, the operator disclaimed any responsibility for his employee's actions. The recruiter, almost always a Negro, had to exercise cunning and guile to evade detection, for to be caught stealing labor could mean death or at least a severe whipping. Many of the itinerant Negro preachers who went from camp to camp were actually recruiters in disguise. They persuaded workers to slip away and accept employment elsewhere by promising higher pay and better living conditions. Some of the recruiters were paid a fixed sum for each man enticed into their employer's camp. Later the same recruiter might receive pay for taking the same workers to the employment of another.[1]

The system of recruitment often worked against the Negro. John Gary, a Negro of some education, was sent to Georgia by his Alabama employer in 1897 to obtain about a dozen workers. Gary found twelve Negroes willing to come to Alabama, but when the tickets that had been sent to the Georgia railway agent were requested, the agent refused to release them. Gary, though aware that the tickets had arrived, did not press the agent for them, well knowing that to do so might mean bodily harm for him. Instead he marked time in Georgia for a few days and then walked all the way back to Alabama. One operator who ruled his Negroes with an iron hand had refused to permit any communication through the mails with the world beyond his camp at times when labor was scarce. In addition he opened all letters sent to his camp, removing any money enclosed which might have enabled the Negroes to leave him. On one occasion a group of Negroes en route to Mississippi were forcibly removed from the train by their former employer, who claimed that they had slipped away before settling their accounts with him.

The system of credit provided the most effective means by which the operators were able to maintain a supply of labor. In every turpentine camp the owners had a commissary where the Negroes drew supplies and provisions against the money payments due them for their labor. After 1900 most operators issued coupons to their workers good only at the company store. In many localities, cash payments for labor were made only once each year, usually a few days before the Christmas holidays. In emergencies workers could obtain small cash loans from their employers.[2]

Many of the operators pursued a deliberate policy of keeping their employees in debt by encouraging them to buy articles that they did not need and could not afford. Overcharging the Negro for what he bought and dishonestly running up a bill for goods not purchased were common procedures. Since most of the workers could neither read nor write and had to depend upon the employer to keep the records correctly, they could not protect themselves. Many of the Negroes lived with no thought of the future and purchased goods in as great an amount as their employers would allow. To the average Negro, being in debt was of little consequence, for always there would be pines to chip and his employer to provide him with a hut and food for his subsistence. It was only when the Negro decided to seek employment elsewhere that he faced the ill consequence of debt.[3]

Since many of the Negroes were drifters, always searching for better working and living conditions, only the fact of being in debt prevented them from moving to new locations. The scarcity of labor in Mississippi around 1900 caused many of the operators to use any possible means to retain workers. It was a general practice among turpentine men forcibly to bring back those who had left while in debt and to compel them through labor to pay off their financial obligations. Dennis Smith, an old turpentine worker, remembered when Negroes with a debt of only fifty cents had been pursued with bloodhounds and brought back to work out their debts. According to Pete Fairley, another old turpentine worker, when a Negro slipped away owing a debt, the hue and cry was out for him and he was returned regardless of the distance, if found.[4]

The employer often sought the aid of law enforcement officers in apprehending Negroes who were in debt. Rewards were posted for runaway Negroes and descriptions of them were sent to the localities where they were expected to go. In many instances the runaway Negroes were protected by one operator from another, particularly if the new employer was short of labor. Nevertheless, it was extremely difficult for the Negro to escape his debts since they gave the operator a lien on the labor of the worker. Later, after naval stores became concentrated in the hands of the large operators, the Negro's debts were transferred with him when he moved to another locality.[5]

The turpentine villages or quarters where the workers lived were usually the private domains of the owners, and their rules accordingly embodied the full authority of the law. It was not unusual for the powers of life and death to be exercised over the Negroes by the overseer. If the superintendent or overseer was humane, as many of them were, the Negroes fared well. If the boss was cruel and overbearing toward those living in his jurisdiction, however, the lot of the worker was often unbearable. The boss or superintendent of the village settled all the differences among Negroes and between them and himself without recourse to law. Indeed, the intervention of the law in matters concerning Negroes was not welcome to the employer, for to imprison a worker for a petty crime meant the loss of his labor. Only in instances where the boss was unable to handle a situation was an officer of the law called in.

Often when Negroes were jailed for petty violations, their employer got them out of jail, paid their fines, and sent them back to work in the woods. Most operators preferred to run their villages without the assistance of the law and considered themselves to be responsible to no one in meting out justice to their Negroes.[6]

As many of the bosses were little better than outlaws, the deaths of Negroes from violence were seldom reported to the press or brought before the courts. In some instances the death of one Negro at the hand of another was unknown beyond the confines of his village. In such cases no penalty was exacted for the crime, for it was the policy of the operator to bury the dead and save the living, especially if the survivor was a good chipper. But if a white man was killed by a Negro, the full penalty of the law was exacted, often without the formality of a judicial trial. Conversely, the killing of a Negro by a white man went almost unnoticed.[7]

In the period between 1895 and 1908 many of the Negroes who migrated to Mississippi were criminals and escaped convicts. As is usual in a new country, the lawless element in the population was perhaps large. All the operators and woods-riders carried pistols; doubtless a few were forced to kill in self-defense. In some instances in Harrison and Jackson counties, on the other hand, the bosses were said to have murdered Negroes with little provocation. The camp of one operator located in Harrison County was named the "graveyard" by the Negroes because of rumors that many of its inhabitants had disappeared and were never seen again. Another camp in Hancock County bore such a dark reputation that it was avoided by the turpentine workers. But many operators exercised their overpowering authority with moderation, pursuing the policies best calculated to obtain the maximum output of labor at the least possible cost.[8]

One observer reports that the turpentiner was almost a nomad. About every five years a given orchard or plantation was exhausted and the still, store, Negro huts, and all equipment were moved to a new location. Usually the Negroes followed the operator from place to place and depended upon him for supplies, counsel, and protection. Some of the turpentine operators had in their employment Negroes whose fathers and grandfathers were the slaves of their fathers and grandfathers. Life in the turpentine camps was often rougher and more primitive than in logging camps. Yet, according to one observer, some of the operators carried on their establishments in the spirit of the old South.[9]

J. A. Simpson, superintendent of the Finkbine Lumber Company Naval Stores, maintained that by giving the Negroes a fair deal he encountered little if any difficulty. A stern taskmaster, Simpson compelled the workers to be out of the quarters during each workday of the week. His camp at D'Lo had a colored Y.M.C.A., Negro baseball teams, fraternal orders, and a church. Leonard O'Neill, who was superintendent of the L. N. Dantzler Lumber Company naval stores business, also had little trouble with the Negroes. He detected troublemakers quickly and got rid of them. According to O'Neill, the average turpentine worker was ignorant, good-natured, and childlike, and he looked to others above him to make decisions even concerning his personal life. Gary, an old turpentine worker, agreed with O'Neill that the turpentine Negro was ignorant, uninformed, and incapable of managing his affairs. But his condition, in Gary's opinion, was the responsibility of the white men who often directed him with brutality, refused to give him schools, and subjected him to a system of ruthless economic exploitation. Dennis Smith, another turpentine worker, stated that the production of naval stores was "outlaw work carried on by outlaws." It was outlaw work in the sense that those who chipped the boxes and gathered the crude gum were rarely paid more than a bare subsistence wage. In the sense that the work was superintended by outlaws Smith thought that the bosses were inhuman and devoid of human decency, for only about one out of every four of his employers had treated him humanely. As he saw it, the ill treatment suffered by those of his own race had caused them to become ignorant, inconsiderate, insensitive, and unkind to each other.

Most of the turpentine workers lived in shacks grouped closely together, known collectively as the "quarters." These houses were built in rows, with the front of each facing the street. Small plots of ground were set aside next to each cabin to enable the occupants to grow their own vegetables. Many of the cabins built in Mississippi around the turn of the century were one-room huts, made of pine poles and possessing neither floors, doors, nor windows. The ordinary dwelling had one room, a front porch, and a stick and dirt fireplace. If the family were large, a second room, known as the lean-to, was added to the structure. Houses of the workers were built to serve a temporary need. When the operators moved to a new location, the simply-constructed buildings were pulled down and moved to a new site. Most of the Negro families lived in the quarters, preferring the communal life characteristic of villages to the social isolation that came with living out in the forest far apart from a neighbor.

Life in the quarters, especially in the larger ones, provided for the basic social needs of the workers. Often the first building to be constructed in a new location was a house of worship. Religion among the Negroes was primitive and highly emotional. The Baptist and Methodist denominations predominated. The preacher, almost always a worker, was one of the most respected members of the colored community. Nearly all the Negroes were members of the church which served as an important outlet for their deep spiritual nature.

Although it is difficult to generalize concerning the wide variation in the social customs and practices of the Negroes, one may safely say that few of the turpentine Negroes attached any importance to the marriage ceremony. They ordinarily selected their mates without benefit of clergy or a civil marriage. In one large turpentine camp, marriage was arranged simply by the couple's asking the boss for living quarters. The overseer placed a stove, a bed, and a table in a vacant hut, and asked the man if he would accept the woman for a wife, and the woman if she would accept the man for a husband. If both answered affirmatively, they were, to all appearances, married. Though marriage without a license was not lawful, the Negroes respected it and considered it binding. Divorce, equally easy, was also granted by the employer. Removal of the furniture from the building ended the marriage, and man and woman were again free to select new mates. Many of the so-called marriages were temporary liaisons that ended when either of the couple moved to a new location. Illegitimacy among the turpentine people bore no moral stigma, and infidelity was of no special concern. Competition for the favors of a particular man or woman, however, sometimes resulted in scrapes involving the use of razors by the women and knives and pistols by the men.

J. A. Simpson, believing that a marriage ceremony would

tend to make the Negro families more enduring, and in turn create a more stable supply of labor, in 1914 ordered all Negro families employed by him to secure a marriage license and go through a legal ceremony. The Negroes were given a definite time limit within which to comply with the order or to leave the village. Few of them refused. Couples who had lived together for thirty years and were parents of grown children had their union legally confirmed. According to Simpson, the Negroes derived a great deal of pleasure in going through the formality of the marriage ceremony.

Few Negroes were able to resist games of chance. The Georgia shin game played with conventional cards appeared to have been the best loved; even craps and taking a chance with the dice were less common. The turpentine operators, aware of the Negro's propensity to gambling, usually constructed a building a short distance from the quarters where games of chance would go on undisturbed night and day until the few lucky ones had garnered all the money. Professional Negro gamblers called "hustlers" met payday at every camp. Occasionally the white boss put his camp out of bounds to the hustler, who might, if caught, be severely whipped. To the knifing and shooting affairs which often accompanied gambling no one paid much attention. In at least one instance a turpentine operator skillful at cards was able to win back a part of the sums that he had paid out to his workers. In general few operators attempted to prevent gambling, and it was accepted by most as a common weakness of the turpentine Negro.

Along with gambling, almost all Negroes were addicted to alcohol and tobacco. In most cases the bosses recognized their inability to prevent Negroes from indulging in drunken orgies and refrained from interference. Generally, if the Negro was physically able to perform the work assigned to him, no effort was made to prevent his drinking. Often, however, when excessive drinking created disturbances in the quarters, offenders were punished. In a few camps consumption of alcoholic beverages by the workers was encouraged by the boss through commissary sales.

From the early days to about 1905, the Negro was issued a weekly ration in addition to an occasional money payment consisting of a peck of unbolted corn meal, four pounds of smoked pork, and a quart of syrup. By 1905 the practice of giving rations was discontinued, largely because money payments were being made more frequently. In addition to the coarse foods obtained at the company store, the Negroes supplemented their meager diets by catching wild animals in the forest and gathering green herbs in the woods. Squirrels, 'possums, 'coons, skunks, rabbits, and turtles were especially relished by the workers. Almost every Negro was an expert fisherman. Tender buds of the palmetto were said to have a flavor similar to cabbage. Young tender bamboo twigs, according to Dennis Smith, tasted like snap beans.

Usually the Negro turpentine worker prepared his three meals in the evening after the workday had ended. Supper, his heaviest meal, ordinarily consisted of meat, syrup, and cornbread. About four-thirty in the morning he got out of bed and made off to the forest. At eight or nine o'clock he ate a light meal. The Negro carried a hollow reed with him to suck water caught in the turpentine boxes to quench his thirst. Rarely did he drink from the streams, for their waters were believed to be the cause of fevers.

As in the case of food, much of the medicine for the treatment of common physical ailments came from the woods. A medicine made from the leaves of the boneset plant was supposed to cure vomiting if the leaves were stripped up the stem. Medicine from the leaves stripped down the stem would act as a purgative. The coin-shaped leaves of the dollarleaf plant were brewed into a tea that was supposed to check dysentery. To treat flux, a disease apparently common among turpentine Negroes, a tea was made from boiling the leaves of the yellowtop plant. A remedy for fever was the tea derived from boiled black snakeroot. Butterfly weed was believed to be effective for the treatment of diarrhea. Tea made from boiling smart-weeds was used to cure rashes, itches, burns, and other skin ailments. In all the large camps free medical service became common. It is probable that malarial fevers, colds, and venereal diseases were the most common ailments among turpentine people.

Superstitions about disease and the belief in magic or luck was almost universal among the Negroes. Witchcraft practiced by Negroes known as witch doctors or voodoo men seems to have been general in some localities. Lather gathered on the face of a sleeping man was a sign that a witch was riding him. When witches were angry with people they would get out of their skin and ride upon the backs of people. Certain objects or articles were supposed to have the power of transmitting the indefinable some-thing called luck. The tobay, a good luck charm, consisted of a lodestone sprinkled over with graveyard dust and covered by a piece of red cloth. The tobay was supposed to give one a winning hand of cards, or to cause a woman to give her favor to the possessor of the charm. The jo-mo, similar to the tobay, was another object from which the possessor might expect good luck. It was thought that charms would not work unless obtained from certain people possessed of the power to impart luck to the object. Negroes with physical peculiarities were usually considered to be in touch with the spirit world, and they became the vendors of charms. Witch doctors and other merchants of charms made their way from camp to camp plying their trade. Some of them, like a few of the Negro preachers, were recruiters; if their true identity was discovered, they met short shrift. On occasion it was dangerous for a witch doctor to sell a charm that failed. One itinerant doctor sold to a swain a charm charged with the power to make a certain village belle succumb to his advances. But when the magic of the witch doctor had no effect, the enraged suitor, believing himself defrauded, with poetic justice slew the purveyor of the love potion.[10]

Southern conservatism in techniques of production and marketing of naval stores was matched in the fields of labor supervision and management. Apparently the labor policies of the industry were unparalleled anywhere in the industrial system, except on the plantation in the black belt. The turpentine Negro was illiterate, ignorant, unambitious, diseased, skilled only in the crude simple tools of his trade, and almost wholly subject to the whims and caprices of his employer. There was and is no pressure group to give voice to his special needs. As a result, the social security system covering almost every worker has only recently been extended to include the turpentine worker. The few hundred naval stores Negroes continue to live in cabins not greatly dissimilar to those of their ancestors, and now as then remain largely isolated in an eddy untouched by the swift-moving currents of modern times.

CHAPTER XI

LUMBERING REACHES ITS PEAK

AROUND THE TURN OF THE CENTURY, the pioneer phase of the lumber industry in Mississippi came to an end. Development of transportation facilities and markets and the flow of capital to the longleaf pine country ushered in an era of large-scale lumber production. Lumbermen from the old timber districts of the North and East migrated to the South in considerable numbers between 1890 and 1915. These newcomers, attracted by the last great body of virgin pine timber east of the Rocky Mountains, brought capital and managerial experience with them which contributed greatly to the development of the yellow pine industry in the Lower South. In Mississippi, a not inconsiderable number of native lumbermen also became large operators during this period.

The proximity of the yellow pine country to areas of highest lumber consumption was a powerful stimulus to the exploitation of southern softwoods. Compared with the great virgin forests of the Pacific Coast, those of the

Gulf states were almost two thousand miles nearer to the consumers of the Northern and Eastern states. Moreover, the distances to Western Europe, Africa, and to the eastern shores of South America were much less from ports on the South Atlantic and Gulf of Mexico than from the Pacific Coast harbors.

Another factor of equal importance in the development of the yellow pine industry was the existence of millions of acres of relatively cheap timberlands, for the most part untouched by man. Northern and eastern lumbermen were amazed at the sight of longleaf forests on flat and rolling land, almost devoid of undergrowth, or trees of other species, stretching for hundreds of miles. Moreover, stumpage could be purchased cheaply prior to 1905, and sawmilling in the warm South was a year-round operation unhampered by snow and ice.

R. A. Long of Kansas City, owner of large blocks of southern timber and many sawmills, in 1902 predicted excellent returns for investments in southern pine:

I believe that I can truthfully and correctly say that no great body of timber has ever made or promises to make as good a per cent of profit for its investors as has yellow pine. As to beauty of growth, in my opinion, there is no other forest under the canopy of heaven that can compare with it.[1]

In 1900 Robert Fullerton, president of the Chicago Lumber and Coal Company, a large manufacturing and wholesale firm, stated:

Of the various kinds of lumber today, yellow pine holds the most prominent position and finds its way into the most remote parts of the world. This has all been accomplished within comparatively few years. The demand is on the increase and unless we make an effort to husband our timber supply, we will find ourselves within a very few years regretting our imprudence in foolishly making haste to grow rich by manufacturing into lumber pine trees which, if allowed to stand, would repay us threefold.[2]

The development of the Mississippi lumber industry had been comparatively slow until the late nineties. In 1880 there had been 295 sawmills with a total capital investment of slightly less than $1,000,000. Ten years later capital invested in 338 mills had risen to $3,092,684. In 1899 the capital investment reached $10,800,000 and 608 mills were in operation, the combined output of which was slightly more than 1,000,000,000 board feet of lumber. During the next five years capital poured into the state in unprecedented volume. In 1904 the total investment reached $24,819,000 and the value of the annual products, $26,162,000. In 1906, 184 new mills were erected and only five went out of commission, to give the state an additional 500,000 board feet of daily mill capacity. In 1909 total capital investment stood at $39,455,000, number of sawmills at 1,647, and the value of output at $42,793,000. However, the number of mills, production, and value of timber products were less in 1914 than 1909, a decrease which may have resulted in part from a period of depression.[3]

During the period between 1900 and 1909, Mississippi emerged as one of the leading lumber producing states. This fact is demonstrated by the following statistics:

Year	Production in Board Feet	Rank Among States
1900	1,202,334,000	—
1904	—	8
1905	1,299,390,000	8
1906	1,840,250,000	5

1907	2,094,485,000	4
1908	1,861,016,000	3
1909	2,572,669,000	3
1910	2,122,205,000	3
1911	2,041,615,000	3
1912	2,381,898,000	3
1913	2,610,581,000	3
1914	2,228,966,000	3
1915	2,200,000,000	3

The increase in production of lumber in the state was believed by some at the time to be reaching its peak in the years 1907-1913. George S. Gardiner stated in 1907, for example, that few large mills were being established and that some were even going out of business. While large blocks of timber were still held by investment companies, Gardiner nevertheless believed that output of lumber would increase but little in the future. J. E. Rhodes also believed that the longleaf pine industry was fully developed in 1913, and that the climax of production had been reached.[4]

Longleaf pine lumber manufactured in the southern part of the state constituted the bulk of the Mississippi lumber output. Out of a statewide cut of 1,299,390,000 board feet in 1905, 1,107,191,000 board feet were longleaf pine. In 1906 only 500,000,000 board feet in a total cut of 1,840,250,000 board feet were not longleaf. J. F. Wilder, a Mississippi lumber manufacturer, reported that 82 per cent of the 1909 total cut of 2,600,000,000 board feet was longleaf pine.

Data on the number and location of mills illustrate still further the rapid expansion of the lumber industry before 1907. Adjacent to the railroads, hundreds of small mills were erected between 1890 and 1905; and near them villages, towns, and even a few small cities grew up overnight. Many of these mills were erected along the Gulf and Ship Island Railroad. In 1897, when the railroad was no more than sixty-five miles in length, it served eighteen sawmills; three years later the number had grown to nearly sixty mills. In 1901 the cut of the mills adjacent to the road had reached 300,000,000 board feet. In January-October, 1902, 27,879 freight cars of lumber moved over the road; during the same months in 1903, 29,014 cars. The lumber industry on the Gulf and Ship Island and its branch lines continued to expand during the next few years, with the total daily output of the mills in 1906 estimated at 4,500,000 board feet. In 1907 about one-tenth of all the yellow pine lumber manufactured in the Southern states, or about 800,000,000 board feet, was transported by the Gulf and Ship Island line.

The picture was much the same in regard to the Mobile, Jackson and Kansas City Railroad, a system constructed in the years 1896-1903 from Mobile northwestward through the entire length of the longleaf pine country of east Mississippi. South of Newton, the northern boundary of the longleaf country, there were sixty sawmills along its route in 1905 with an output of 1,500,000 board feet daily. The lumber industry served by the Illinois Central had almost reached its peak by 1902. About thirty-five mills located on the road had a combined annual production of 250,000,000 board feet. In 1901 only four new mills were built, though output was much greater than in the previous year. Apparently the greater output resulted from an increase in the size of mill units.

The New Orleans and Northeastern Railroad, known later as the Southern, passed through a magnificent longleaf forest between the shores of Lake Pontchartrain and

Meridian. From the time of its completion in the middle eighties to 1906, forest industries developed rapidly along its route. In 1900 the combined production of the mills located on the New Orleans and Northeastern was 300,000,000 board feet.

The lumber industry on the Mobile and Ohio Railroad remained small in comparison with that along the other railroads in the pine country of Mississippi. With its northern terminus at St. Louis, the Mobile and Ohio was, however, one of the major carriers of lumber from the South. Shipments of lumber over the road to St. Louis were 18,924 carloads in 1901; 24,574 in 1902; 23,272 in 1903; 20,754 in 1904; and 24,883 in 1905.

Total shipments of lumber from Mississippi by rail were a good measure of the lumber industry in the interior. In 1902-1903 the average volume of rail traffic was about 12,300 carloads per month, or approximately 147,600 carloads for a twelve-month period. Of this amount about 118,000 cars containing 1,500,000,000 board feet went to domestic markets, and about 29,600 cars to the export trade.[5]

Although, as we shall see, large mills eventually came to dominate the Mississippi lumber industry, mills sawing from 5,000 to 15,000 board feet, nevertheless, accounted for a large percentage of the lumber output before 1909. When a railroad was constructed through a virgin forest, numerous small mills sprang up as though by magic. Being unable to compete in the long run with the big operator in acquisition of superior timberlands or in efficiency of manufacturing, many of these small mills either closed down or moved to new locations after the timber belonging to homesteaders and lesser landowners located near the road had been cut. Many of them, however, were able to eke out an existence by utilizing isolated and inferior tracts which were uneconomical for the large millmen to exploit. As the sawing of materials for building railroad cars required no special equipment, this job was especially well suited for the small millman. On a favorable market he could profitably turn out car material even though his product could not command the price the big manufacturers obtained for a similar grade. Thus it happened that in periods of prosperity, usually when car material was in high demand, hundreds of small mills went into operation almost overnight, only to be closed by the next slight recession.

Even before 1890 the trend was in the direction of ever larger manufacturing units. In 1902 the *American Lumberman* reported that men of capital and conservative interests were beginning to dominate the southern lumber industry. Experienced and forward-looking lumbermen were making investments of $100,000 or more in this region. In 1906 the editor of the same journal reported that there were then more than twenty mills with a daily production of 60,000 board feet and more in the southern longleaf pine country. Two years later some of the larger operators were producing from 35,000,000 to 200,000,000 board feet annually.[6]

Several factors favored the establishment of large mills in the Lower South. The manufacture of lumber was a complicated process requiring many and comparatively highly paid workers whose services the small millman usually could not afford. The large manufacturing unit by using labor-saving machinery and specialization also was able to produce lumber of superior quality and at lower cost than was possible for the small unit. Yet there was an optimum size. A mill of 500,000 feet daily capacity was regarded as uneconomical, because such a mill, if it were to be kept in operation over a long period, would require the transportation of timber over excessive distances. Moreover, the costs of depreciation were high, and moving a large mill from one location to another was impracticable. Probably a mill of 70,000 to 200,000 board feet capacity, built to operate in one location from fifteen to twenty years, was the unit of optimum size in the longleaf pine country.

Conditions in the longleaf section of Mississippi were remarkably favorable for the large unit. In the first place, this region possessed an immense body of timber of superior quality concentrated in a large geographical area. In the second place, title to most of the virgin timberlands had been acquired by a few owners. Concentration of ownership made it easy for millmen to buy acreages in contiguous blocks sufficiently large to justify the expenditure for mill and other equipment. The extent to which consolidation of timber ownership had progressed is illustrated by the fact that in 1903 two-thirds of the timberland in Mississippi was owned by nonoperators and one-third of that by only six persons. The remaining one-third was distributed among seven mill owners.

A mill cutting 30,000,000 board feet annually needed about 20,000 acres of virgin timberland to supply a ten-years' run. The cost of such an establishment was large even in the nineties and increased enormously in the ensuing years. It was estimated in 1898 that a double band mill would cost $175,000; the 20,000 acres of timberland, $60,000 to $90,000. In 1907 the same mill would have cost $325,000 and the timberland $1,200,000. In 1915 the initial cost of a plant of average size built to operate for fifteen years was $500,000, and the value of its timber supply was five to six times the value of the plant.

In the large manufacturing units of the longleaf pine country, three different types of saws, the circular, gang, and band, were in use. Gradually the band saw supplanted the circular and gang saws in mills that catered to the domestic market, and was also widely used by the mills producing lumber for export. After stumpage prices became high (1900-1912) the band saw, which was less wasteful, proved superior to other types of saws.[7]

In the big mill the initial sawing operation was usually accomplished on the second floor of a rectangular two-story building. The logs were conveyed from the mill pond to the sawing room by an endless chain. As a log entered the sawing room, a large swinging circular saw cut it into the desired lengths. When the sawyer removed the stop loader, the log rolled upon the carriage, where it was fastened by setting works consisting of headblocks and dogs. In the sawing process, which was controlled by the sawyer with the assistance of carriage men, a log was fed against the saw at maximum capacity. In the first step, the outside parts of a log were removed, leaving a square stick of timber which was subsequently divided into smaller pieces. Only a few seconds were needed to saw up a log, and mental calculations of great speed were demanded of those concerned in the work.

When a large output was desired, the headsaws, performing the initial sawing operations, were used only to divide the log into pieces of large widths called cants. These cants were then moved on chains to resaw machinery, located a short distance to the rear of the headsaws, where the sawing operation was completed.

Near the resaw machines was the edging machine. Its principal functions were to square the end of the boards, divide lumber into different widths, and separate the superior pieces from the inferior ones. The edgerman sought to cut out the knots and to eliminate, if possible, the other defects in the board which might lower the grade. To run an edging machine required a thorough knowledge of lumber grades, alertness and quick thinking, great physical strength, and endurance.

From the edging machines lumber went either to the dry kiln or to the green yards. The function of the many different types of dry kilns was to dry or season the lumber so that it would retain permanently its shape and color. Usually the kilns were of smaller daily capacity than the mills, and a large percentage of the lumber was air-dried in the green yards. The time required to dry lumber in kilns varied from thirty to ninety hours depending upon the grade of lumber. (See Plate I, Figure 3.)

Most of the dry-kilned lumber went to the remanufacturing plant commonly known as the planing mill. There the lumber was planed, smoothed, and shaped. Matching machines, surfacers, moulders, sizers, and other mechanical devices were used to produce the finished pieces. Moulders were designed to finish flooring, ceiling, and drop siding at high speed. Another machine, the cutter, planed off the coarse grain to produce smooth-surfaced boards. Sizers ordinarily were used for smoothing large pieces.

Almost all lumbermen were compelled to build large lumber yards in which to store their products when orders were slack. Many mills had a storage capacity of more than 25,000,000 board feet. In the early years lumber was handled by hand, a time-consuming and expensive method. As labor became short, millmen turned to more economical methods. Endless chains, gasoline motors, and devices for loading freight cars were used to decrease labor costs.

The power required for operating the large number of complicated machines was enormous. All but a few mills in Mississippi derived their power from large horizontal steam boilers which used waste products as fuel. For a mill with a daily capacity of 100,000 feet, about 1,000 horsepower was necessary.

Maintaining mill operations at a high peak of efficiency necessitated establishing blacksmith and machine shops and foundries. The Finkbine Lumber Company, for example, was equipped in its machine shop and foundry to build a locomotive or manufacture any part required for mill operation. Large outlays of capital went into the stocking of considerable supplies of spare parts.

Outside the South millmen commonly obtained their raw material from contract loggers, but in Mississippi, contract logging, except on rivers, accounted for only a small percentage of the log supply used by the manufacturers. Never very extensive before 1890, it all but disappeared by 1915. Only where timber was scattered or located in small patches did an insignificant amount of contract logging continue.

Logging in Mississippi was a concern of the large mill operator. When timber was exhausted near the mills, lumbermen constructed railroads known as tramroads (in Mississippi often called dummy lines) to their timber stands. These logging roads were often of standard gauge, and some of them ultimately became common carriers hauling both freight and passengers. By 1905 timber for all the large interior mills, and for many of those with a daily capacity between 25,000 and 60,000 feet, was transported over such tramroads. Many of the lines were more than thirty miles in length, and an average of about six spurs per square mile extended out into the woods. After the timber was removed from the area, the spurs were taken up and relaid into uncut timber. The cost of spur construction per thousand board feet of timber transported was sixty cents for one firm in 1911, and for three firms, in 1915, from 27 cents to 48.4 cents.[8]

The amount of equipment used in logging depended upon the consumption rate of the manufacturing unit and the distance that logs had to be transported from woods to mill. To supply logs to a mill of approximately 100,000 board feet capacity required between forty and fifty flat cars and from one- to three-rod engines known as shays. The shays were low-speed, sturdily built locomotives designed to pull cars over the temporary spurs. Heavy geared locomotives known as main liners pulled from ten to forty carloads of logs from a central assembling point in the woods to the mills. (See Plate I, Figure 2; Plate VI.)

Although coming somewhat later than in the mills, labor-saving machinery was standard equipment in most logging operations by 1908. About 1900-1905 the ancient caralog for hauling logs was supplanted by the eight-wheel log wagon (see Plate VII), which in turn was superseded in most large operations by steam skidders and log loaders. Small operators and a few large lumbermen who were concerned with protecting young unmarketable timber refused to use skidders in their timberlands. Some lumbermen clung to the old methods on the theory that the cost of buying and maintaining skidders was so high that they did not actually decrease logging expenses. Many large millmen turned, however, to the mechanization of logging equipment both because it made them less dependent on the unstable labor market and because it permitted logging to go on uninterrupted by the rainy seasons so prevalent in the coast country.[9]

Skidders were used on a small scale as early as 1893 in Michigan and the yellow pine district of North Carolina. Their first use in Mississippi was probably in the logging operations of the J. J. Newman Lumber Company near Hattiesburg in 1894.

The skidder and loader combination was becoming known in the longleaf pine country in 1901. An editorial plug in the *St. Louis Lumberman* in 1902 stated that logging by steam had come to stay. Any outfit cutting from 50,000 to 150,000 feet daily could decrease logging costs from fifty cents to $1 per thousand by using steam skidders. Another advantage in skidders was that by their use logging operations in low woodlands could go on regardless of heavy rains.

The skidders were set up near the spur tracks where timber was thick. The steel wire cables, with varying lengths up to 1,000 feet or more, were unwound from the drums and attached to the logs in the woods. As the revolving drums reeled in the cables, four or five logs or more were skidded to the spur track at one pull-in. Each Lidgerwood four-line ground tree-skidder of the Great Southern Lumber Company pulled in 519 trees of forty- to eighty-foot lengths, about 200,000 board feet in a normal work day. The Finkline Lumber Company used a Clyde skidder that skidded all the logs on four acres at one setting; and on each pull-in, five to fifteen logs were snaked to the track.[10] (See Plate VIII.)

But logging with skidders brought complete destruction of young timber of unmarketable size. (See Plate IX.) No trees or vegetation of any kind except coarse wire grass remained on the skidder-logged hill and ridges. For miles and miles the landscape presented a picture of bare open land that graphically illustrated the work of destruction wrought by the economic activities of man. Except for a lone misshapen tree here and there, the rolling hills and flat lands appeared to be a treeless country destitute of fertility. Nor was the work of destruction a temporary condition, for twenty-five years later the boundaries between skidder-logged areas and those where other methods prevailed were apparent even to the untutored eye. In the skidder areas, scrub oak and inferior hardwoods formed a mantle over the land preventing the return of longleaf pine. Where conservative logging practices prevailed, vigorous young forests sprang up to cover the land once more.

The lumber industry reached its peak growth in the years 1908-1914. At that time, almost all the timber was in the hands of a few manufacturing companies which would continue operations until the virgin forests were gone. To a remarkable degree, production of lumber in the longleaf pine country came to be concentrated in the hands of a few large producers. The existence of millions of acres of virgin timber in contiguous blocks, a plentiful supply of capital, and the large demand for timber products were the basic factors that brought concentration in the lumber industry. A less tangible factor was the general trend toward larger and larger productive units which became common in almost all types of American industry. That concentration did not go further than it did in the lumber industry was due to the wide geographical distribution of the raw materials and the decentralization of ownership.

CHAPTER XII

THE BIG MILLS

As we have seen, by 1915 the Mississippi longleaf pine industry was dominated by a relatively few large producers. Along the coast sawmilling had passed its peak. In the interior the big mill phase of lumbering continued to be vigorous until the late twenties, but only a few large-scale operations existed in the early thirties.

At the mouth of the Pascagoula River, the firms operating with an annual production of between 15,000,000 and 40,000,000 board feet were Hunter-Benn, Will Farnsworth, Wyatt Griffin, and J. Bounds.[1] Only the L. N. Dantzler Lumber Company, which had two mills at Moss Point and others elsewhere, could be classified as a large-scale manufacturer. At the mouth of the Pearl were the old medium-sized firms of H. Weston and Poitevent and Favre. In 1913 the Poitevent and Favre Lumber Company moved a few miles westward to Mandeville, located on the northern shore of Lake Pontchartrain. The Ingram-

Day Lumber Company operated a large plant at Lyman, a few miles north of the coast.[2]

More is known about the history of the L. N. Dantzler Lumber Company than any other. The origin of the company can be traced to 1849, when William Griffin with others developed a lumber business at the confluence of the East Pascagoula and Escatawpa rivers. This small antebellum lumber firm was dissolved in 1860, when Griffin became the sole owner. Following the Civil War, Griffin resumed mill operations in a lumber business that was continued without interruption by members of his family down to 1938.

Lorenzo Nolley Dantzler became connected with the Griffin mill business as a result of his marriage, in 1858, to Erin Griffin, the daughter of William Griffin. Dantzler, born in 1833 in Perry County, Mississippi, was the son of John Lewis Dantzler, who migrated from South Carolina to the Leaf River bottoms in Mississippi in 1808. L. N. Dantzler attended Centenary College and afterwards served in the Confederate Army during the first years of the war. As the conflict drew to a close, Dantzler, while a purchasing agent for the Confederate Government, was imprisoned by Federal authorities for a short time on Ship Island.

After the war, Dantzler became a partner in the William Griffin Lumber Company, which at that time included the elder Griffin and two of his sons. The partnership was dissolved in 1873 when Erasmus Griffin sold his interest for $15,000. Dantzler exchanged his interest in the firm for a small mill, and with it began the long-lived L. N. Dantzler Lumber Company.

Little information survives about Dantzler's early mill business. He then was only one among a considerable number of small millmen operating at the mouth of the Pascagoula. At this time Emile De Smet, A. C. Danner, and the W. Denny Company were the largest producers in the Pascagoula-Moss Point district. By 1880 Dantzler, however, had enlarged the scope of his operations and had acquired another small sawmill and a shingle mill. In partnership with others he also operated a shipyard, brick kiln, towing business, and a sash and blind factory in the years 1873-1880.

Expansion of the firm during the eighties resulted directly from business relations between Dantzler and Henry Buddig, a New Orleans lumber merchant. Buddig encouraged Dantzler to build a large mill, offering to finance the entire cost. With much misgiving Dantzler accepted the offer, and began construction of one of the first large double circular mills at the mouth of the Pascagoula in 1883. Completed in 1885, the mill was one of the most modern in the coast country.

The erection of the new mill was completed at a critical time. The demand for export lumber and timber was growing, and Dantzler's competitors at the mouth of the Pascagoula were not slow to realize that a new era in lumber production, characterized generally by mass production, was rapidly approaching. The increased demand for lumber, in addition to the proximity of millions of acres of first class timberlands to rivers emptying into Pascagoula Bay, made the Moss Point-Pascagoula district an important center for the export trade.

The continued expansion of Dantzler's operations during the nineties resulted from the success of his business dealings with upcountry logmen. Like other coast mills, the Dantzler Company purchased the greater part of its timber from logmen who were financed by the company. As a condition of loans made to the timbermen, the company took out a life insurance policy on the borrower as well as a mortgage on his property. When a logman's returns were less than the amount of the loan, his timberland was claimed by the company in settlement of his debt. In a number of cases logmen were furnished money to buy state lands which passed quickly into the company's possession.

Acquisition of large tracts of timberland along the Pascagoula and its tributaries made it possible for the Dantzlers to eliminate their competitors in the Moss Point-Pascagoula district. The firm gradually became the largest purchaser of timber and logs, thus gaining a near-monopoly in rafting on the Pascagoula. The Dantzlers' virtual control of raw materials after the turn of the century caused one millman after another to cease operations, until in 1910 only three fairly important concerns remained.

The Dantzlers foresaw the tremendous expansion of the lumber industry. In the late nineties L. N. Dantzler, Sr. was lamenting the prospect that his age would preclude his playing an important part in the great expansion soon to come. His second son, John L., though educated for a career other than in business, was in fact to be the architect of the company's growth. His three other sons, who also were active in the business, were reaching maturity strategically on the eve of the new era. But it was the characteristically optimistic secretary, financier and general manager, John Lewis, who proved to be the driving force behind the firm's expansion. According to his contemporaries John Lewis was a sharp trader, close-mouthed, and possessed an unusual ability to remember minute details.

The first important phase of company expansion occurred immediately following the construction of the Gulf and Ship Island Railroad, which ran through the general area where the Dantzlers had previously acquired considerable tracts of timber from upcountry logmen. From Howison and Rogers the company in 1899 purchased a recently completed mill of 70,000 board feet daily capacity and a few thousand acres of virgin timberland. The newly acquired circular mill, which was designed to cut timbers for the export trade, the virgin timber nearby, and a railroad to Gulfport were the company's chief inducements to extend its manufacturing enterprises into the interior. Two years later they bought a mill located at Handsboro and timberlands belonging to Henry Leinhard. The old gang mill at Moss Point in operation before the Civil War was now replaced by a new gang and circular mill. The resulting increase of mill capacity brought company production to 90,000,000 board feet annually. The firm had become one of the largest lumber manufacturers in Mississippi.

Unusual conditions initiated the second important period of expansion. In September, 1906, a devastating hurricane swept over the coast counties, blowing down from one-third to two-thirds of the timber in some localities. R. Q. Breland, land agent for the Dantzlers, estimated their loss of standing timber at close to $1,000,000. Time proved, however, that such an estimate of storm damage was too high, since most of the downed timber, if sawn before becoming infested with worms, could be saved. Much of the leveled longleaf and slash pine timber was unaffected either by weather or worms.

The Dantzlers moved rapidly after the storm, erecting a number of portable mills to salvage their downed timber as well as that of others which they purchased for only a fraction of the going price. They constructed a large storage area at the mouth of the Pascagoula River and brought storm timber in large quantities down the streams. The high prices of lumber then prevalent, coupled with the low price of this timber, enabled the company to reap large profits from this salvage operation.

In the next year the lumber industry was overtaken by a severe panic which continued over a fairly long period. While most domestic producers curtailed their production, the Dantzlers, possessing excellent connections abroad, for a long period continued to run their mills night and day. Indeed, the average price of lumber received by the Dantzlers, $17 per thousand, was exceptional, though prices of export lumber of superior quality cannot properly be compared with the average quality and lower prices of lumber sold on the domestic market. Under such conditions their profitable operations in 1907 and in 1909-1910 encouraged the Dantzlers to undertake further expansion.

In 1910 the Dantzlers agreed, for a commission of $1 per thousand board feet, to operate the large millplant of the defunct Bond Lumber Company. After five years of successful operation, they purchased the Bond Lumber Company mill which possessed a capacity of 100,000 board feet daily. In the same year they purchased the Ten Mile Lumber Company properties also, which included a circular mill of 80,000 board feet capacity, 30,000 acres of timberland, tramroads, a commissary, and a mill village. With the addition of the Ten Mile mill the Dantzler firm became the largest lumber manufacturer on the Gulf and Ship Island between Hattiesburg and Gulfport. In partnership with Edward Hines the company erected a large mill at Kiln. Another mill owned jointly by the Dantzlers and the Standard Export Company was acquired at Piave.

The company reached its peak of growth in 1913. Value

of mills, timber, lands, and other properties then was estimated at $3,372,691.40. In view of the fact that the company owned between 400,000 and 500,000 acres of land, the estimate appears conservative. Assets representing unpaid accounts and accumulated surplus amounted to $5,472,612.

Obviously the Dantzlers anticipated the end of large-scale lumber manufacturing in Mississippi, for in order to continue operations they acquired large timber holdings elsewhere. In 1913 they acquired important timber concessions in Nicaragua, although they erected no mills there chiefly because of the prevailing political instability in that country. They also purchased four hundred shares in the Sumpter Lumber Company, a Florida concern. On Prince Edward Island the company obtained a timber concession but built no mills. In partnership with the H. Weston Company the Dantzlers bought timberland in Oregon with a view to continuing lumber manufacturing there, but that project eventually was abandoned.

The Dantzlers, like other large lumber concerns, combined both production and distribution. As in the case of other coast millmen, the Dantzlers shipped almost all of their lumber to overseas markets in Europe, South America, the Caribbean Islands, and Mexico. Large quantities of lumber were sent in company-owned schooners to Cuba and other Caribbean Islands, and to Mexico. Hewn and sawn timber went mostly to England and European countries, but the great proportion of lumber was marketed in South America.

With their production high, the firm organized an export company in partnership with Price and Pierce, international timber merchants. Lumber sent to Europe and South America was marketed through the Standard Export Company, but that sold to Mexico, Cuba, and islands in the Caribbean was distributed through company agents. By 1907 the company had become one of the largest shippers of lumber in the state. In that year the firm shipped 145,368,376 board feet at a price of $17.80 per thousand. Thirty-five cargoes containing 45,484,057 feet were sent to Buenos Aires, while another thirty ships loaded with timber and lumber went to Europe. During the period 1907-1915, the year 1913 was the peak year, marking the shipment of 149,000,000 board feet. The price, the highest yet received, was, on an average, $18 per thousand. Of the total amount sold only 10,333,980 board feet were disposed of on the domestic market.

The prosperous year of 1913 was followed by a sharp reduction in demand associated with the beginning of World War I. Matters became even worse when the lucrative Latin American market was dislocated by the European conflict. In consequence all purchases from other lumber firms were stopped and production in company-owned mills in 1914 was reduced to 62,000,000 board feet. Most of the mills were out of commission in 1915, when output sank to 49,000,000 board feet.

The history of the L. N. Dantzler Lumber Company after 1915 is fairly typical of other large producers in the longleaf pine belt. The preparedness program immediately preceding America's entrance into World War I inaugurated a period of unparalleled prosperity. In 1918 the average lumber prices obtained by the company were higher than $25 per thousand feet. The following year, despite the payment of a 10 per cent dividend on its stock, there was an earned surplus of $7,669,970.

The Dantzler Lumber Company, one of the oldest firms in the state, remained in business after almost all of the other large concerns had ceased operations. As their supply of virgin timber shrank, one mill after another was stilled, until finally in the thirties only one plant continued to operate. The last log sawed at the Dantzler Moss Point mill in 1938 ended the manufacturing of lumber at a center which had been the scene of lumbering for almost a century. It also spelled the end of the big mill period which had had an early start at the mouth of the Pascagoula River. The Dantzlers ceased to be manufacturers, but retained 115,000 acres of cutover land which has since been converted into a tree farm. So ended the largest and one of the oldest lumber manufacturing firms in the coast country.[3]

In the interior section of Mississippi west of the Pearl River, large sections of timber had been cut long before 1900. For this reason lumbering enterprises after the turn of the century were smaller than those in the virgin timber country east of the river. Indeed, there were only four fairly large operations in the western sector during the early 1900's: the Pearl River Lumber Company, with an annual capacity of more than 40,000,000 board feet; J. J. White at McComb; Enochs brothers at Fernwood; and the Butterfield Lumber Company at Norfield. Each of the last three had a capacity ranging between 15,000,000 and 40,000,000 board feet.[4]

The Butterfield Lumber Company, a family-owned enterprise, typified the medium-sized lumber business found in the interior. Though not among the oldest firms, it was one of the first, if not the first, to acquire a modern band sawmill. In 1887 J. S. Butterfield and Frank Norwood bought two small mills and a few years later built a double band sawmill near Brookhaven. When Norwood disposed of his interest to Butterfield in 1896, the entire mill business became the property of the Butterfield family. The Butterfield mill in 1907 had an annual capacity of 30,000,000 board feet and was capable of manufacturing lumber in both long and short lengths.

The town of Norfield grew up near the location of the Butterfield mill. Like others, Butterfield built houses for his workers and produced all the services needed by his employees. Unlike most other millmen, however, Butterfield built no camps for loggers in the forest near the scene of timber cutting. In fact, the logging methods employed by the firm were different from those used by other firms. For example, Butterfield developed the Rooney logging cart after much experimentation and employed it in his business. This vehicle, drawn by mules, had two low wheels with wide tread, and though of small capacity possessed the important attribute of maneuverability. In addition, the company was among the first to discontinue contract logging, because the contractors habitually cut only the best timber, leaving the rest to expensive logging operations later.

The Butterfields owned one of the finest virgin longleaf pine tracts in the South. In 1890 the company had obtained 20,000 acres of timberland, and this with other tracts acquired before 1907 brought its total holdings to 45,000 acres. The average volume of the Butterfield virgin stands ranged from 12,500 to 15,000 board feet per acre. In texture and uniformity of growth the timber was unequalled by any other tracts of similar size in Mississippi.

In 1915, the Denkmann-Reimers firm, owned principally by the F. C. Denkmann Lumber Company, purchased from the Butterfields the sawmill at Norfield, 450,000,000 board feet of stumpage, and the Natchez, Columbia, and Mobile Railroad. The Denkmanns were also owners of large lumber interests in other parts of the state and in east Louisiana.

With the sole exception of the Mobile and Ohio, the railroads constructed east of the Pearl ran through virgin forest. Large blocks of almost untouched forests encouraged the concentration of lumber production in large units.

Fairly large operations were begun immediately after the building of the railroads; towns sprang up almost overnight. Bond, Estabutchie, Hillsdale, Poplarville, Purvis, Wiggins, and many others owed their beginning at this time to the booming lumber industry. The two most important lumber centers east of the Pearl River, however, were Hattiesburg and Laurel. Hattiesburg, a small village in 1885, was located at the geographical center of the longleaf pine country between the Pearl and Pascagoula rivers. It became important in both distribution and production of lumber. In 1893 there were fifty-nine sawmills with a daily output of 1,000,000 board feet in the Hattiesburg vicinity. One observer thus prophesied the growth of Hattiesburg:

> being located about the middle of the Gulf and Ship Island and with its ninety mills, the terminus of the Mississippi Central, with about twenty-six mills, and about midway between New Orleans and Meridian on the Northeastern and the line of Mobile, Jackson and Kansas City, with possibly thirty mills, it is easy to see that the future of Hattiesburg is assured.[5]

In 1905 almost every wholesale concern in the United States had representatives in the city, and 6,000,000 board feet of lumber moved through there each day. The population of Hattiesburg was estimated at 21,000 in 1906 and four years later at 25,000.

Thirty miles northeast of Hattiesburg the New Orleans
and Northeastern Railroad produced another lumber manufacturing center. Following the construction of the railroad, a small sawmill was erected by John Kemper at the
point later named Laurel. The location selected by Kemper
was in the approximate center of a large virgin forest and at
the intersection of two important railroads. After 1895
four large mills were erected. In 1915, when the daily
output of the Laurel mills reached 750,000 board feet, the
city was the largest producing center in the state.

Between 1900 and 1915, lumber operations of medium
size were common in all the longleaf pine counties east of
the Pearl. Most of the enterprises of this category owned
considerably less than 100,000 acres of timberland, but few
held title to less than 20,000 acres. Among the millmen
whose annual output ranged from 15,000,000 to 40,000,000
board feet were A. G. Little, at Richardson; Pine Burr
Lumber Company, at Pine Burr, owned by F. V. B. Price;
Progress Lumber Company at Stoner; Headly Lumber
Company, at Poplarville; and the Pole Stock Lumber Company, owned by Herbert Camp, at Hattiesburg. At Hillsdale was located the mill of the Southern Land and Timber
Company, owned by Rand Batson and the Hatten brothers,
Norman and Wade. Other firms were the Tatum Lumber
Company, owned by the Tatum family at Bonhomie; the
mill of Emmett McInns of Hattiesburg; Butler McLana-
han's plant at Estabutchie; P. M. Ikler's at Moselle; and
John Hinton's at Lumberton. Other millmen were Mul-
ford Parker, who had one mill at Kola and another at
Ellisville, and William Conner at Seminary. The Silver-
thorne Lumber Company conducted mill operations at
Williamsburg, and H. H. Crust owned a mill at nearby
Arbo. Other medium-size mills were the K. C. Lumber

Company, owned by Gregory Luce and others at Lucedale; the J. M. Griffin mill at Nomac; Richton Lumber Company at Richton; George Leatherbury's mill at Bexely. B. D. Moore, Butler McLanahan, H. S. Hagerty, and others owned mills at Benmore and Avery. George Robinson's mill at Chicora was the only medium-size mill on the Mobile and Ohio.

There were eleven large firms in the interior district east of the Pearl. Most of the concerns there continued in the lumber business down to the late twenties, when the virgin forests were depleted. Such firms as the J. J. Newman Lumber Company at Hattiesburg, Edward Hines with mills at Lumberton, Kiln and elsewhere, the Finkbine Lumber Company with mills at D'Lo and Wiggins, and Eastman-Gardiner at Laurel owned large tracts of timberland. The Marathon and Wausau lumber companies at Laurel, owned by W. H. Bissell and others, were among the largest producers, although they held less than 150,000 acres between them.[6] While the exact extent of its land holding is unknown, the F. C. Denkmann firm with mills at Ellisville, Ora, Brookhaven, Mish, Norfield, and others in east Louisiana, was one of the largest producers in Mississippi and the South. The Chicago Lumber and Coal Company, primarily a wholesale and retail firm, operated a large mill at Lumberton. Gilchrist and Fordney's mill at Laurel had an output of about 60,000,000 board feet annually. The Mississippi Lumber Company, owned by Pearly Lowe and others, operated a large mill at Quitman and owned large acreages of timberland.

The Eastman-Gardiner firm was one of the few large family-owned lumber operations in the interior of the pine country and one of the first of the northern lumber companies to come south. Stimson Gardiner, founder of the company, was born in New York and moved first to Pennsylvania and later to Iowa. The Gardiners, including George and Silas, sons of Stimson, became associated with the Lamb-Byng Lumber Company of Clinton, Iowa. In 1890 the operations of the Gardiners at Clinton were in their last stages because of the growing shortage of white pine timber.

In the late eighties the Gardiners began to buy timber in Jones County, Mississippi. They purchased a small mill in 1890 and soon after began the construction of a larger plant at Laurel. In 1899, when other large plants were being erected, the Gardiners increased the annual productive capacity of their Laurel mill to 60,000,000 board feet. The company also built a large cotton mill to utilize mill waste as fuel. By 1903 the company had acquired about 180,000 acres of land located mainly in Jones, Jasper, and Smith counties.

The average selling price of lumber at the mill in 1902-1903, according to George S. Gardiner, was $11.77 per thousand, while the total cost of production was $10.50 per thousand. The company's capital investment exclusive of timberland was between $600,000 and $700,000. In 1903, the firm shipped 3,530 carloads of lumber.[7]

The Eastman-Gardiner Lumber Company was one of the larger yellow pine firms of the state. In 1904 it employed 1,000 workers with an average weekly payroll of $35,000. Its logging facilities were extensive, consisting of seven locomotives, eighteen miles of spur tracks, and 125 logging cars.

By 1904 the J. J. Newman Company had become the largest lumber manufacturing enterprise in the interior of the state. The development of the firm was representative of the consolidation of mills and timberland which had become prevalent in the pine country. The principal owner, Fenwick Peck, was born at Scranton, Pennsylvania, in 1854. Peck's father had established a profitable mill business at Elmhurst. In 1885 the younger Peck organized the Lackawanna Lumber Company, a $200,000 corporation, and acquired three sawmills. In 1892 the firm's capital was increased to $750,000, and a large tract of timber was purchased.

Realizing that the South was destined to become a major lumbering section, Peck came to Mississippi in 1896 and traveled through the pine forests in the Hattiesburg vicinity. He eventually joined forces with the J. J. Newman Lumber Company, which had purchased the Wiscasset mill in 1892 and was operating what was termed a first-class yellow pine mill at the time of Peck's arrival. The Pennsylvanian increased the firm's capital, and at a later date bought 147,000 acres of timber from Knapp and Stout. Additional land acquisitions in the Hattiesburg vicinity brought total holdings of the company to 400,000 acres.

In 1901 Peck's far-flung lumber interests in Pennsylvania, New Mexico, and Mississippi were merged to form the United States Lumber Company. At the same time the annual capacity of the Hattiesburg plant was increased to 75,000,000 board feet. A few months later construction of two other large mills got underway at Sumrall, a few miles west of Hattiesburg. In 1907 the Mississippi Central Railroad, which Peck owned, was completed from Hattiesburg to Natchez. Peck continued to expand his operations until the production of his mills in Mississippi reached 200,000,000 board feet annually. In order to dispose of this large output, the firm established a wholesale lumber concern. With others, Peck also established the Homochitto Lumber Company at Bude on the Mississippi Central

Railroad. In 1905 the total investment of the United States Lumber Company in timberlands, sawmills, and in a railroad in Mississippi was estimated at $26,000,000. The company played an important role in the economic development of that part of the state where its operations were located. Indeed, much of the growth of Hattiesburg can be attributed to the Peck mill in the town and to others situated nearby. On the Mississippi Central west of Hattiesburg there were about twenty-six mills in early 1905.

The Edward Hines Lumber Company, another large corporate enterprise, acquired tremendous holdings in the area between the New Orleans and Northeastern and the Gulf and Ship Island railroads. Edward Hines, founder of the company, started out in the wholesale lumber business and expanded rapidly. In order to insure his Chicago wholesale and retail business a stable and continuous supply of lumber, Hines acquired both timberlands and sawmills of his own. In this period of expansion into lumber manufacturing, Frederick Weyerhaeuser became an important stockholder in the Hines Lumber Company.[8] Hines was among the first of the lumber wholesalers to carry large stocks of yellow pine. Foreseeing the end of his supply of northern white pine lumber, he began to purchase longleaf pinelands in Mississippi and eventually acquired between 240,000 and 300,000 acres. Hines, in 1906, bought his first manufacturing interest at Ovisburg which included a sawmill, tramroad, flat cars, a mill village, and a turpentine distillery. In partnership with the Dantzlers Hines built a large mill at Kiln and eventually became the sole owner. When the Mississippi virgin forests were depleted, the Hines Lumber Company left the state and moved to the Pacific Coast.

The era of big mills, 1890 to 1930, though of short duration, represented a distinct stage in the economic history of the longleaf pine section. Millmen borrowed and invested heavily to erect the mills which all too quickly consumed the virgin forests. During these years the competitive advantage of large mills eliminated many of the smaller operators and placed ownership of all the factors of production in the hands of a very few lumbermen. The improvement in manufacturing techniques employed in big mill operations paralleled the same trend in other industries. Wherever practicable the lumber manufacturer replaced men with machines. Mass production and specialization of labor was emphasized in the struggle to achieve lower production costs. As the southern lumber industry approached maturity greater and greater aggregations of capital were required to erect the type of manufacturing unit in general use. In fact, after 1900 few men starting with little capital were able to become large producers.

CHAPTER XIII

MISSISSIPPI PINE GOES ABROAD

T HE GREATER PART of the southern pine lumber produced on the Mississippi Coast was consumed in foreign countries. In fact, for a long time yellow pine was better known in Western Europe than in the great consuming sections of northern United States. Even before the Civil War square timber, ship masts, planks, and other timber products had been exported from the Gulf to Europe, Australia, the West Indies, and Latin America. These markets were largely responsible for the development of the first large manufacturing units on the Gulf Coast during the 1880's. Subsequently the foreign markets continued to absorb an important percentage of pitch pine lumber and timber even after a heavy demand had developed in the United States. At the beginning of the twentieth century, more pitch or yellow pine was consumed in Germany than all other woods combined. In addition, longleaf and slash pine were especially favored by the Germans for manufacture of window casings, wagons, freight cars, and other uses which exposed wood to the outdoors. In 1902 the pitch pine of superior quality produced on the Pascagoula and Pearl rivers virtually supplied the prime lumber market in Europe.

Around the turn of the century approximately 800,000,000 board feet of lumber were shipped annually to foreign markets from Mobile, Pensacola, and Pascagoula. During 1904-1905, 2,250,000,000 board feet, not including hewn timber, were exported from Gulf ports, a larger quantity by approximately 500,000,000 board feet than had been shipped in any previous two-year period or in any five-year period of the decade 1890-1900. In the business year ending July 30, 1913, 1,332,683,000 board feet went from Pensacola, Mobile, Gulfport, and New Orleans, the four largest lumber exporting ports in the world.

To a certain extent southern lumbermen were free to choose between the foreign and the domestic markets. However, the foreign market generally demanded a superior quality of lumber and a great variety of grades. Thus the millman producing small amounts of inferior lumber was excluded from the export trade. Many operators followed the practice of sending their best grades overseas, while marketing their inferior grades in the United States. Philip Gardiner asserted in 1910 that nearly all the southern manufacturers depended upon the foreign market for disposal of a portion of their products. Many of the mills located south and east of Hattiesburg sawed either wholly or in part for the export trade. In 1913 most of the large concerns were exporting an important percentage of their production to the overseas market. Such millmen as Poitevent and Favre, H. Weston, George Robinson, Farnsworth, Wyatt Griffin, W. Denny, and the L. N. Dantzler Lumber Company habitually sold almost their entire production to the overseas market. In the years 1907-1915 the L. N. Dantzler Lumber Company marketed 794,662,000 board feet, all but 40,000,000 feet overseas.

Lumber produced at the mouths of the Pearl and Pascagoula rivers and along interior railroads was loaded aboard ship at Gulfport. The neighboring town of Pascagoula was also an important lumber port. From July 1, 1900, to June 30, 1901, 111,584,536 board feet were shipped from Pascagoula to overseas countries. In 1906, after the Pascagoula harbor had been deepened, ships with a carrying capacity of 1,000,000 board feet took on cargoes of lumber at the mills. Yet Gulfport outdistanced her sister port as a lumber export center because of the twenty-two foot channel dredged from the mainland to Ship Island, and also because it was the southern terminus of the Gulf and Ship Island Railroad.[1]

In the harbor at Gulfport a large storage area was constructed for the assembly of sawn and hewn timber awaiting shipment during the winter months. During the summer season such timber was stored in Bayou Bernard, a short distance northeast of Gulfport, to keep it away from the teredo worm, a timber destroyer present in the salty waters of the Gulf during warm weather. From the storage place rafts of timber were towed out to Horn and Ship islands where the lumber was loaded aboard ship.[2]

After the harbor was opened at Gulfport, as many as forty vessels, steamers, brigs, barks, and schooners from all parts of the world might be found taking on cargoes of lumber. In loading, heavy timbers were shoved through openings in the hulls of schooners or other sailing vessels by means of rollers. In the hulls of lumber-carrying sailing vessels there were a number of compartments arranged one above the other. As a ship gained in weight with each stick of lumber loaded, it sank deeper, allowing water to flow into the open lower compartment. After being filled with lumber, the compartment was sealed, and then another above the water line was opened.

Marketing lumber abroad was vastly different from selling in the United States domestic market. Because of tariffs on remanufactured lumber, heavy timbers and lumber in a rough state were exported to some countries. Export lumber was generally cut to grades specified by terms of agreement between buyer and seller. Sometimes large export firms contracted for the entire cut of a number of mills for a given period. In one case, Southerland and Innes handled the total output of the Rose Lumber Company and others. In another, the Reeves-Powell Company contracted for a time the export cut of the L. N. Dantzler Lumber Company, Denny and Company, Moss Point Lumber Company, Poitevent and Favre, and the H. Weston Company; and for six months the Standard Export Company handled the entire cut of the Great Southern Lumber Company. United States shippers and exporters often sent shipments to agents overseas who advanced from 80 to 90 per cent of the value of the cargo to the manufacturer and then sold it for a commission. This practice was known as "selling direct." Exporters also bought lumber from manufacturers and contracted for delivery to foreign firms at a specified time.[3]

Arbitration was an essential feature of every contract between an exporter and a foreign buyer. In substance, the usual contract provided that lumber should conform as nearly as possible to the specifications of grade and quality agreed upon by the two contracting parties. The buyer was obligated to accept the shipping documents before the cargo reached him, leaving the question of damage and violations of the agreement to be adjusted by arbitration. This was a frequent source of misunderstanding between sellers and buyers. The American seller all too often was represented in arbitration proceedings in foreign countries by agents over whom he had no effective control and who were not always disinterested parties in the negotiations.[4]

"Consignment shipping" was sending lumber to an overseas agent who sold the lumber for whatever price was offered. In this system of marketing, the shipper had no control over the sale price of his product; upon occasion it failed even to bear the freight and insurance costs. After American pitch pine had gained a world-wide reputation, lumbermen believed that consignment sales were unnecessary and that they depressed the price of lumber.

As the scale of their operations increased, southern millmen sought through the establishment of export companies of their own to eliminate one of the steps between themselves and foreign merchants. When they entered the manufacturing end of the business, exporters and timber merchants tried to do virtually the same thing. Price and Pierce, British exporters of pitch pine, owned over six million acres of timberland in Canada beside two mills in the Moss Point-Pascagoula district. The Standard Export Company, owned partially by the L. N. Dantzler Lumber Company, was one of the largest yellow pine producers in Mississippi. In 1904 the number of combination manufacturing and export firms shipping lumber from Gulfport about equalled the number which exported only.

The trend toward combination of producers and exporters failed to eliminate the exporters. They continued to serve the needs of many large producers and especially of small millmen. The average lumberman generally needed the immediate returns to be gained by selling to the exporter instead of waiting for the slower returns from sales overseas. As the carrying capacity of ships increased, the cut of several mills was required to assemble the variety of grades and quality of lumber needed to complete a cargo. Only large operators cutting a large variety of grades were able to assume completely the functions of exporters.

Export manufacturers considered the absence of a uniform classification of lumber a serious handicap to themselves. Each country had its own peculiar building customs which required lumber grades unlike those of other countries. Hence, sawing lumber for customers in a number of countries was complex, time-consuming, and wasteful. Since in most instances only superior grades of lumber were acceptable in the foreign markets, the problem of disposing of inferior grades, some 15 to 60 per cent of each log, was perplexing. Before 1890 only the best portion of the tree was manufactured into lumber, the remainder being wasted. When prime timber grew scarce, however, manufacturers attempted to establish a system of grade classification which would enable them to use inferior timber and still maintain the price for first class lumber.

The problems of price, grade classification, and inspection continued to plague the manufacturers from the nineties down to 1915. From time to time lumbermen tried to control the market through industry-wide co-operation, but they met with little success. Because of the diversity of ownership characteristic of the industry, a trade association offered the only hope of achieving control of production and prices; therefore, the Gulf Coast Yellow Pine Exporters Association was established in 1892 to fix prices and to create a uniform lumber grade classification. Its membership included thirty-seven coastal mills and a number of others on the New Orleans and Northeastern Railroad as far north as Hillsdale and on the Illinois Central between Lake Pontchartrain and Bogue Chitto. The members adopted a lumber grade and price classification. George W. Robinson, a Moss Point millman, was appointed to negotiate with Atlantic Coast producers to gain their acceptance of the Gulf Coast lumber rules. This exporters' association proved to be short-lived, however, and its objectives were never reached.[5]

A new organization, the Gulf Coast Lumber Company, including several of the larger manufacturers, was organized in 1895. It was intended to control the output of all the mills owning stock in the company. According to its regulations, members of the company who sold their products below the fixed price were to be penalized. Five directors represented the coast mills of Alabama and Florida and three those of the Mississippi Coast. In 1895 and 1896 the company issued a lumber classification list and in 1896 effected a curtailment of production reported to have increased average lumber prices from $9 to $10 per thousand board feet. Again, however, co-operation failed to develop. Competing nonmembers shaded their prices sufficiently to undersell the company and thus upset the price scale. Failure had been inevitable, for 40 per cent of the export producers refused to join the organization. In the important Moss Point-Pascagoula district, for example, only one millman belonged to the organization.

During the last years of the nineteenth century, the prices of hewn and sawn timber were depressed. At 10½ cents to 12 cents per cubic foot in 1897, they were slightly lower than they had been in 1892. It was chiefly because of the prevailing low prices that lumbermen tried again to better their lot through co-operation. A group of millmen meeting at Mobile in 1897 agreed to raise their lumber prices. At this meeting George Robinson recommended that an organization be set up for the purpose of processing and distributing trade information to lumbermen.[6] Out of Robinson's ideas grew the Export Bureau of Information which had objectives similar to those of earlier export-manufacturers associations. Prices on a number of grades of lumber were advanced successfully by the association. In 1900 the association was reorganized and renamed Gulf Coast Lumbermen. Once again the association members failed to secure the co-operation of a number of the important producers, one reason being that each export district was competing with the others. How successful the organization actually was in influencing prices is difficult to gauge. In 1899 lumber prices did advance rapidly. During a short period kiln-dried saps went from $9 to $15 per thousand board feet, while prime lumber rose from $16 to $20, and sawn timber from about 12 cents per cubic foot to 15½ cents. Nevertheless, the association soon was dissolved.[7]

The association fell victim to an improvement in trade conditions. Production was involuntarily curtailed in 1901 and 1902 by a long drought which prevented the mills from getting their usual supply of logs by water. As a result, demand was high early in 1902, and it was difficult for the coast mills to fill the orders. Mills having a supply of logs operated both day and night. Prices continued to rise throughout 1902: prime lumber sold on the overseas market in the later part of the year for average prices of $22 and up per thousand board feet and sawn timber brought from 14 to 18½ cents per cubic foot. In 1903 and 1904 the upward trend of prices continued, with the demand for pitch pine particularly strong in Argentina.

In 1905 both the demand and the prospect for yet higher prices were good. At that time prime lumber was selling for $30 per thousand board feet. The year 1906 brought still higher prices; lumbermen experienced the most prosperous season in their history. Sawn timber prices passed twenty-seven cents per cubic foot. While prices were at such unprecedented levels, southern lumbermen had few grievances and less desire to co-operate with each other.[8]

The heavy demand for lumber and the resulting high prices continued well into 1907. In January of that year prime lumber brought $35 per thousand board feet at Mobile. The lumber manufactured in the interior was diverted to the foreign trade by an acute car shortage. By February, however, the market showed signs of weakening, and in June both prices and demand were much lower than in the previous month. Although prices continued to fall, export mills for a time were less drastically affected by the panic than were the domestic ones. In September, November, and December, the Pascagoula mills were still operating, and one of the largest concerns even kept its plants running night and day. While export conditions were less favorable in 1908 than in 1906, they were, nevertheless, better than those prevailing in the domestic market. The prices received for sawn timber were in some instances as much as eight cents per cubic foot lower than comparable 1906 prices. In the latter part of 1908 the fluctuating prices assumed an upward trend, with prime face eleven-inch lumber bringing $32.50 per thousand board feet. Apparently the depression did not seriously affect the operations of the Dantzler Lumber Company, for in March its mills operated around the clock. In 1909 the prices and demand, while much less satisfactory than in 1906, nevertheless compared favorably with the lows of 1907 and 1908.

Conditions in the export trade were none too satisfactory either for manufacturers or exporters in 1910 and 1911, partly because prime timber was becoming scarcer with each passing year. Once more, despite the repeated failures of such efforts in the past, export lumbermen formed an association for the purpose of establishing uniform grade classifications for their products. A second objective of the association was to devise an equitable method of settling claims made by foreign buyers against Gulf Coast exporters and millmen. According to Hugo Forcheimer, who was an exporter, reclamation payments of this kind averaged about $1 per thousand board feet or $1,500,000 annually.[9]

Disagreement among the members of the new association immediately developed over the problem of establishing uniform grade classifications. Exporters and millmen from Pensacola desired a classification founded upon a percentage basis; that is, a specified percentage of the lumber would conform to a certain grade. On the other hand, Gulfport millmen and exporters wanted all lumber to conform exactly to the grade specified in an order. Upon this problem the association members could not reach an agreement. Most of the manufacturers favored the position of the Pensacola exporters. However, failure to establish a uniform grade classification was due also to opposition by some exporters. Because they were dealing with a number of countries where different grades of lumber were required, some exporters apparently believed that uniformity of grades was neither desirable nor possible. When lumbermen, who outnumbered exporters in the association, pushed through a grade classification despite opposition from the exporters, the latter withdrew from the association. Without these members the association soon ceased to exist.

In respect to claims and reclamations of foreign buyers, the only solution, in the opinion of some contemporary observers, was rigid inspection of lumber at its point of origin. Differences between American and foreign methods of conducting business were at the root of the difficulty. Time was valuable in the United States, and quality was regarded as secondary to quantity. In a single afternoon a cargo of 60,000 board feet might be loaded and shipped with only a cursory inspection at Gulf ports. When the ship arrived at its destination on foreign shores, however, each piece of lumber was minutely examined, and large quantities were rejected as substandard in quality. This experience led foreign buyers to demand stricter inspection of lumber before it was shipped and uniform grade classifications which could make the work of the examiners easier. Throughout the period the purchasers set the terms of sales, and reclamations against American producers remained numerous.

The demand for lumber was stronger throughout 1912 than it had been during 1909-1911, and the domestic producers turned a part of their output to the overseas market early in 1912. The trend to higher prices continued unchanged during the first half of 1913. In the second half of the year heavy consignment shipping brought a reversal of the trend. Foreign buyers, having bought high and now being forced to sell on a low market, were caught in a price squeeze. In order to recoup some of their losses, they resorted to the expedient of filing excessive claims against the American exporters and manufacturers.

The outbreak of war in 1914 further accentuated the depressed condition of the market. In August of that year all except two of the L. N. Dantzler Lumber Company mills were closed down. Only a trickle of lumber shipments from the Gulf Coast was reaching Europe. European demand decreased by 58 per cent or more over a period of a few months, and prices declined below production costs.[10] During the first few weeks of the war it was uncertain whether trade with Latin America would also be affected. Certainly the South American trade would be embarrassed by the fact that most of the lumber sent to that market was sold through European firms.[11]

After a few months the needs of war created strong markets for American lumber in France and Britain. One French order for 50,000,000 board feet was divided among several Mississippi and east Louisiana firms. Exporters and shippers were unable to charter steam vessels because of their scarcity and the high insurance rates. In consequence, the old sailing vessels were brought back into use to transport lumber overseas from ports on the Gulf of Mexico.

Foreign countries purchased the bulk of their supply of softwoods from the Southern states. Longleaf and slash timber because of their versatility and low cost met little competition from other softwoods. The existence of the foreign market provided the initial stimulus for the beginning of the coast lumber industry, and markets abroad continued to absorb large quantities of Mississippi-produced lumber. In fact, the full development of lumbering in Mississippi would have been long delayed had not an overseas market existed in which lumber and timber could be sold for prices profitable to the millman.

The channels of trade between the Mississippi producers and the overseas purchasers were complex and often unsatisfactory to both. Much of the difficulty was due to the dissimilarity of American and foreign business methods. The Mississippi lumberman, wedded to mass production, was prone to strive for volume rather than quality of product—a practice which ran counter to the desires of the foreign customer. Co-operative efforts of lumbermen designed to impose American standards upon the foreign consumer resulted largely in failure. The dissimilar building customs of different countries made impossible the establishment of uniform grades by the producers.

CHAPTER XIV

MISSISSIPPI PINE
STAYS AT HOME

THE GREATEST PERCENTAGE of lumber produced in the longleaf pine country after 1895 was sold in the states located north of the Ohio and east of the Mississippi River. In 1902-1903 the overseas trade took a total of 29,600 carloads of yellow pine, the domestic market, 118,000 carloads. Before discussing this larger market, let us examine the manner in which the pine lumber was distributed from manufacturer to consumer in the United States.[1]

Wholesalers in the lumber industry in the seventies fixed the grade classification of lumber, contracted the output of the mill at a flat price, and set the price for lumber purchased by the retailers. In time there came to be a number of types of wholesalers with functions roughly similar in nature. The so-called legitimate wholesale dealer was set apart from others who bridged the gap between producers and retailers or consumers by reason of his large investment in his business. Legitimate wholesalers usually owned large lumber yards in the North and East where they sold directly to the retailer and in some instances to the consumer. Their agents in the yellow pine districts bought their supplies of lumber directly from the millmen. At Hattiesburg, a distribution center, thirty important wholesale concerns had purchasing offices in 1911.

Others who called themselves wholesalers, such as commission men, brokers, and jobbers, differed from the legitimate wholesaler in the smallness of their investment in the business. They received lumber on consignment from the manufacturer and disposed of it for a fee. Interested principally in the volume of sales and little concerned with the price of lumber, they did not serve well the interest of the producer.

Near the end of the nineteenth century, there developed a trend toward the integration into one ownership of the various branches of the lumber industry. Many large consumers acquired mills and timberlands. Retailers and wholesalers became manufacturers to obtain a sure source of supply. Manufacturers, particularly the larger ones, developed sales agencies that sold directly to retailers and even to consumers. Thus the Chicago Lumber and Coal Company, the Finkbine Lumber Company, and the J. J. Newman Lumber Company established wholesale offices in distribution centers and in some instances developed line retail yards.

Other millmen tried forming co-operative selling agencies to market the output of a number of mills. Theoretically, this system was the best for the average millman. The agency could offer a large variety of grades, and the sales cost could be distributed among a large number of producers.

Despite the trend toward integration and the experiments with selling agencies, the middleman continued to play an important role in the movement of lumber from the mills to retailers and consumers. The average lumberman did not manufacture a sufficient variety of lumber grades to meet the demands of retailers and consumers. Also his limited operations could not bear the cost of a large sales force. Wholesalers, in close touch with the nation-wide market and possessing the variety of grades needed, could obtain premium prices. In 1911 and in 1915 wholesalers marketed one-half or more of the yellow pine lumber produced in Mississippi.

Problems involved in distribution tended to encourage co-operation among millmen. In the early years of the lumber industry in Mississippi, millmen attempted to control the distribution of their products. Their major aims were a uniform grade, a uniform price, and production control.

In 1888 Alabama, Georgia, and Mississippi lumbermen met at Meridian, Mississippi, to attempt a combination mainly for the purpose of establishing uniform prices. B. E. Brister of Bogue Chitto, Mississippi, called for the co-operation of all lumbermen. He asserted that the best men in all principal industries were uniting to promote their interest. Lumbermen, according to Brister, could prevent the ruinous competition in prices. He deplored the existence of a system "that allowed the ignorant customer to fix the price of a commodity."[2]

Two years later the several local lumbermen's associations, except the coast export producers organizations and others in the Carolinas, Georgia, and Florida, were consolidated in the Southern Lumber Manufacturers' Association. The organization of this regional association reflected the growing importance of the yellow pine industry in the South. Producers catering principally to the export market

and manufacturers of hardwoods took little part in the association. Immediately after its foundation the association began issuing price lists. In 1892 it recommended that its
members curtail production and adhere closely to the price
list promulgated by the association's Committee on Values.
In all fifty-two price lists were issued from 1890 to 1905.

The Panic of 1893 brought overproduction, low prices,
and eventually cessation of mill operations. The association persuaded millmen to limit supply and sought by other
means to establish higher prices. However, the depression
lingered on until 1899, at which time prices started to rise.
The immediate effect of higher prices was competitive
bidding for timberlands and enlargement of mill capacity.
The long-time results were higher manufacturing costs and
greater output potential.

As lumber prices moved steadily upward, lumbermen
were highly optimistic. Through co-operation among themselves lumbermen believed that it was possible to establish
uniform prices that would be profitable to producers. But
the increased mill capacity motivated by the steady up-
swing in prices brought overproduction and decreased demand in 1904.

The state of the market was such that an extraordinary
meeting of the Southern Lumber Manufacturers' Association was held in June to devise a means to restore prosperous conditions. Lumbermen agreed to reduce output for
a two-months' period by one-third of normal production.
Apparently many lumbermen stood by the agreement—70
to 80 per cent of the millmen in Mississippi, Arkansas,
Louisiana, and Texas, whose combined annual output was
2,800,000,000 board feet. Mills eliminated night operations, shortened the length of the work week, or reduced
the number of daily working hours. Reduction of output

in July was estimated at 175,000,000 board feet. In thirty-six Mississippi mills, the July output was 17,144,779 board feet less than average. R. A. Long stated that within ten days after curtailment began the downward trend of prices had been arrested, and that within thirty days values increased. During July the market improved to such an extent that prices were only slightly under the association list. Prices in October were from $1.00 to $1.10 higher than in the previous June.

The market improvement during the latter half of 1904 cannot be attributed entirely to production curtailment. Building permits in twenty-three leading cities in June, 1904, were 42 per cent more numerous than in June, 1903. Prices were rising in the overseas market. The extent of curtailment and output reduction was probably smaller than it was generally represented to be by the lumber journals. In September, 1904, some of the mills operated overtime, while others continued to produce their normal output.

The general prosperity in the United States, in addition to increased foreign demand, brought higher and higher prices until the middle of 1907. Lumbermen in some instances in the years 1904-1907 operated their mills night and day in order to meet the seemingly unlimited demand.

The ever-increasing rise of lumber prices since 1899, a seven-year period, gave the public the impression that the increase was due not to general economic conditions but to price fixing by trade associations. The Southern Lumber Manufacturers' Association, faced with the prospect of investigation and possible prosecution for issuing price lists and engaging in other questionable activities, accepted the advice of attorneys and abolished its price list committee. A new committee known as the Market Committee was established, but it went out of existence in October, 1906. After that time George Smith, secretary of the association, issued market reports from data supplied by sixty correspondents. From November, 1906, to May, 1909, twenty-one such reports were distributed.[3]

In Mississippi, antitrust sentiment was fairly strong. The increase of prices together with the formation of the Mississippi Lumberman's Association (comprising the larger producers in the longleaf section) aroused the suspicion of the governor and some members of the legislature. A House committee in 1906 conducted an investigation of the lumber industry. George Gardiner, testifying before the committee, attributed the increased price of lumber to depletion of white pine forests and to higher cotton prices. He denied that limitation of supply and price recommendations by lumbermen's associations had affected prices. According to Gardiner, 90 per cent of the mills in Mississippi were owned by residents of the state and 95 per cent of the income from the industry remained within the state. Seventy per cent of the gross returns went to labor, and another 20 per cent to taxes and purchase of timberlands.

Partly because of public opinion the name Southern Lumber Manufacturers' Association was changed to Yellow Pine Manufacturers' Association. The organization remained essentially the same in membership and other characteristics, and continued to represent primarily the interests of producers of yellow pine for the domestic market.

The slight break in the market in May, 1907, developed into a full-fledged depression. Late in 1907 men and machinery were idle and production of lumber almost at a standstill. At first the recession was thought to be temporary. Yet weeks passed with little or no improvement.

Curtailment of output, according to the *Lumber Review,* was much more severe in Mississippi than in the other yellow pine producing states. During September, 65 to 70 per cent of the mills on the Mobile, Jackson, and Kansas City Railroad were idle, and the inactivity was not confined to the small concerns. Elsewhere three of the large plants of the Natalbany Lumber Company were silent. Average running time for seventy-two mills in November, 1907, was ten days; for eighty-one mills in December, five days. Early in 1908 only eight of thirty-one mills listed by the Yellow Pine Manufacturers' Association were running. All of the nineteen mills owned by members of the Mississippi Pine Association were idle.[4]

While millmen generally believed that it was better to close down than to saw timber which would, if sold, neither bring a profit nor even defray the cost of manufacture, many of them were unable to cease operating. The prevailing system of financing in the industry was a major obstacle. Mill expansion, acquisition of timberlands, and provision of working capital were to a great extent financed through borrowing by the sale of bonds secured by what amounted to a lien on the standing timber. The millmen paid the taxes on the timber, insured it against the hazards of fire and tornado, and cut it only under conditions stipulated in the trust deed. Brooks-Scanlon, for example, issued $750,000 in 6 per cent bonds on 47,474 acres of Louisiana timberland valued at $3,000,000. Interest payments came to $150,000 or about $1.50 on each thousand feet of lumber manufactured.[5] The necessity of meeting such interest payments and of building up a sinking fund often compelled millmen to maintain operations when returns were little above, or in some instances, even below production costs. For this reason the available supply of lumber might exceed the demand even in a period of extremely low prices. Thus gentlemen's agreements to reduce the supply over an extended period had little chance of success.

The depression and continued low prices motivated lumbermen to seek a new method by which the supply of lumber might be controlled. Early in 1908 active moves were made by many of the larger manufacturers to form a gigantic corporation whereby ownership of mills and timberlands would be consolidated and a wholesale marketing agency established. It was proposed to include in the corporation the major producing firms in the South, whose combined annual output averaged two billion feet or about 20 per cent of the yellow pine manufactured. Those firms which were parties to the proceedings disclaimed any intention of organizing a trust and maintained that their major objective was to conserve timber by limiting the amount harvested. A report showed that investment in mill and timberlands of the prospective members of the projected corporation was $273,000,000, their total indebtedness $42,000,000. Mill and timberland owners in Mississippi who favored the consolidation of the yellow pine industry were F. L. Peck, J. W. Blodgett, F. C. Denkmann, S. H. Fullerton, C. W. Goodyear, Edward Hines, W. A. Gilchrist, and J. N. Stout. It was believed that a three hundred million dollar corporation organized like United States Steel would meet the test of legality. The planners proposed to allocate fifty cents per thousand board feet produced to the United States Forest Service for conservation with the hope of making the consolidation less objectionable in the eyes of the Government. A committee selected to study the proposition reported favorably, and a second permanent committee with R. E. Weyerhaeuser as chairman was appointed to recommend plans for the formation of the corporation. But the move to control both production and marketing through the creation of a gigantic corporation failed when the attorney general of Missouri obtained a court injunction to prevent the merger. At the same time attorneys general in other states took steps to enjoin formation of the company.

In general, the financial panic of 1907 was of short duration, but the lumber industry, except for brief periods, remained depressed until late 1915. Through the Yellow Pine Manufacturers' Association and through informal agreements, lumbermen tried unsuccessfully to limit production and to establish uniform prices. But such efforts could not succeed because productive capacity exceeded consumer demand. For many millmen who had borrowed heavily, limitation of output over an extended period meant financial ruin. In the years 1907-1915 a slight price increase always brought greater production which in a short time created an over-supply and a consequent decline of prices.[6]

It is obvious that the attempts of lumbermen to limit production and to control distribution met with little success. Part of the difficulty can be attributed to the strong antitrust sentiment throughout the country. We have seen how the activities of the Yellow Pine Manufacturers' Association led to the indictment of some of its most prominent members by the State of Missouri. The counts included (1) issuing associational price lists, (2) curtailing output in 1904 and 1908, (3) agreeing to sell lumber only to legitimate retailers, (4) refusing to sell in carload lots to farmer co-operatives, and (5) dividing up the territory among retailers. The lumbermen lost their case and as a condition of doing business in Missouri were forced to resign from the association. So many of the important members were affected that the Yellow Pine Manufacturers' Association received a death blow. In its place the Southern Pine Association, a privately owned corporation, was established in 1914. It made no recommendations concerning prices; its chief objectives were to furnish statistics to lumbermen and to devise a new system of grades.

Unable to establish prosperous conditions through their own efforts, lumbermen appealed in 1915 to the newly created Federal Trade Commission. They appear to have hoped for permission from the Federal Government to establish some system of production control. Charles S. Keith, spokesman for the yellow pine millmen, stated that during the preceding eight years, with the exception of the latter half of 1912 and the first half of 1913, conditions had been unsatisfactory, and had been ruinous since 1914. From $18.42 in 1913, by December, 1914, average price of lumber at the mills had dropped to $11.83 per thousand board feet, while cost of production was not less than $13.75. Edward Hines urged the commission to study the effect of the importation of low grade Canadian lumber on domestic lumber prices.

When white pine was the chief wood sold on the market, it had generally received protection from Canadian lumber. But its growing scarcity on the south side of Lake Huron and Lake Erie prompted some of the white pine producers located near the Canadian border to reverse their position on the tariff. They needed Canadian stumpage that could be rafted easily and cheaply to their mills located inside the American border. These American millmen knew from experience that a tariff on rough Canadian lumber brought in retaliation a Canadian export duty on logs. Such were the reasons for their change of attitude toward the free entry of lumber. American manufacturers had also bought stumpage in Canada as a speculative venture. An abolition of the tariff on lumber would increase the value of their timber holdings.[7]

By 1888 there was some sentiment for either reduction or repeal of the tariff on rough Canadian lumber. One southern lumberman, however, with an eye to the not-too-distant future, predicted that millmen in his section would support protection of lumber. The McKinley Act of 1890 brought a reduction of the duty on low grade lumber from $2 to $1 per thousand board feet. The new schedule would not apply to Canada unless she abolished her export duty on logs. As was expected, Canada repealed duties on logs, but placed restrictions on holders of so-called timber limits unless the owners established lumber mills in Canada. Timber limits were cutting rights acquired by lumbermen on lands owned by the Canadian government.[8]

Southern lumbermen, a new element in tariff controversies, claimed that the duty on rough lumber established by the McKinley Act was too low. Worse still was a bill introduced by Senator William Jennings Bryan in 1892 to repeal the tariff on lumber. The Wilson-Gorman Tariff of 1894 carried out the threat of the Bryan bill by permitting Canadian rough lumber to be imported into the United States free of duty. This act promoted competition between low grade southern and Canadian lumber. According to Edward L. Russell, then president of the Mobile and Ohio Railroad, the Wilson-Gorman Tariff allowed Canadian lumber to flood the country, causing all mills located on his road to cease operations.[9]

Apparently most southern lumbermen believed that free trade in low grade lumber was inimical to their welfare. In 1896 the Southern Lumber Manufacturers' Association recommended restoration of the duties that had prevailed prior to the passage of the Wilson-Gorman Tariff. J. J. White favored a duty of $2 per thousand board feet for rough lumber. A protariff lumberman's convention held at Cincinnati formed an organization to work for protection of lumber.

Lumbermen of the South and those of the Lake states were on opposite sides of the tariff question. White pine manufacturers who wanted cheap Canadian stumpage to operate their mills came out openly for free lumber. Under the McKinley administration, with its strong predilection for high tariff, Congress in 1897 enacted the Dingley Bill, which assessed a two-dollar duty on rough lumber.[10] Canada reacted as was expected by placing an export duty on round, unmanufactured logs. Another retaliatory measure aimed at American owners of Canadian stumpage was an act requiring that logs cut on Crown lands be manufactured into lumber before being exported. This act compelled American millmen who owned stumpage acquired from Ontario to erect mills in Canada.

In the agitation preceding adoption of the Payne-Aldrich Tariff of 1909, pressure was strong from a number of groups favoring abolition of duties on lumber. Conservationists supported tariff revision on the theory that every imported piece of lumber promoted forest conservation in the United States. Consumers believed that high-priced lumber was the direct result of the protective tariff. American owners of Canadian mills and stumpage stood to benefit from the abolition of duties.

Southern lumbermen, almost to a man, were bitterly hostile to tariff reduction on the grounds that Canadian low grade lumber would undersell inferior grades of yellow pine north of the Ohio River. Edward Hines, owner of mills and timber in Canada and Mississippi, led the fight against the low tariff proposal. He asserted that Canadian lumbermen possessed advantages in cheap stumpage and in paying low taxes or none at all on uninhabited land. In Mississippi his annual tax on something over 200,000 acres of timber was $50,000. Canadian lumber, enjoying cheap water transportation, could be brought across the border at a freight rate of $1.75 per thousand board feet, whereas rail shipment of southern pine, which was heavier than Canadian wood, cost from $7 to $10 per thousand board feet to points north of the Ohio River. Figures compiled by Robert Fullerton on freight rates were about the same as those of Hines.[11]

Almost all Mississippi lumbermen were opposed to the free-trade provision on lumber contained in the pending tariff bill. Merchants, bankers, and lumbermen held meetings to devise methods of defeating the measure. J. F. Wilder, president of the Mississippi Lumbermen's Association, contended that tariff reductions would effectively prohibit southern lumber from competing in the low grade lumber market. William Conner, a millman of Seminary, Mississippi, was one of the delegates sent by Mississippi lumbermen to the national capital to fight for a high duty on lumber. Joseph Fordney, a Michigan congressman, part owner of the Gilchrist-Fordney Lumber Company at Laurel, was the only member of the House Ways and Means Committee who desired to retain the duty on low grade lumber.[12]

A lobby committee comprised almost entirely of lumbermen was formed to work in opposition to the free lumber provisions. J. E. Rhodes, secretary of the White Pine Association, apparently led the fight for the lumber interests. He claimed that insurgents in Congress tried to join hands with southern Democrats and the lumber interests for the purpose of overthrowing the regular Republicans. But the lumbermen would have nothing to do with the seceders, preferring to take their chances with regular Republicans. The lumber interests succeeded in getting the House committee to report a one-dollar duty on lumber. But a rule adopted in the House provided for a separate vote on lumber. Proponents of the one-dollar duty felt, according to Rhodes, that the "separate vote" meant that their cause was lost. Rhodes asserted that the lumbermen were desperate and that influences far and wide were brought to bear on Congress. A caucus was held. Trades were made with Democrats covering everything from short-staple cotton to Bibles. Eighteen votes of the Ohio delegation were secured by a trade on wool. Lumbermen won a victory, though not a complete one; the Payne-Aldrich Tariff as passed carried a duty of only $1.25 per thousand board feet on low grade lumber.[13]

With the advent of the Wilson administration and the enactment of the Underwood-Simmons Tariff of 1913, lumber went on the free list. Lumbermen apparently put forth little effort to prevent the change. Their silence did not, however, represent a willing abandonment of position, but rather a realistic acquiescence in the inevitability of a downward revision of the tariff under the new Democratic administration.

In summary of the years 1890-1915, it may be said that for most of the time lumbermen were confronted with low prices, in some instances lower than production costs. The condition resulted from two distinct periods of depression and from over-expansion of the industry. The trade association was established for the purpose of controlling terms of sale. Though at times able to influence the market, manufacturers could never create a uniform price system such as existed in other key industries. Integration of production and distribution in one firm or ownership was general among the larger producers. Independent wholesalers and retailers continued to exist, however, and it was mainly through them that the smaller lumbermen marketed their product. There was a strong tendency toward the consolidation of ownership of both timberlands and manufacturing units in the lumber industry. Such a development did not go further because of the wide geographical area in which lumber was produced, the relatively wide diffusion of timberland ownership, and the antitrust laws of the several states.

CHAPTER XV

RAILROADS
AND THE LUMBER INDUSTRY

T HE DEVELOPMENT OF THE LUMBER INDUSTRY in the interior of Mississippi was almost entirely dependent upon rail transportation. The existence of millions of acres of virgin pine timber provided the initial impulse for the construction of railroads in a section where agriculture was confined chiefly to subsistence farming.

The construction of the first two trunk lines in the late ante-bellum period, the New Orleans, Jackson and Great Northern (later the Illinois Central) and the Mobile and Ohio Railroad, was due principally to the rivalry of Mobile and New Orleans for the commerce of the hinterland. But in this period lumber was not yet thought of: the rich long-leaf and slash pine forests were unbroken and still largely unknown. The building of a network of railroads which traversed almost every district of the pine country was delayed until demand for yellow pine had developed in the great consuming centers of the North and East.

Four of the six main lines in the region were built after

1884, the last one, the New Orleans and Great Northern, being completed in 1909. Feeder roads contructed almost wholly by lumber companies penetrated almost every locality of the pine country by 1910.[1]

The first of the newer main lines, the New Orleans and Northeastern (later the Southern) was built after 1880. It ran from New Orleans in a northeasterly direction through the forests of east Louisiana and Mississippi to Meridian. Captain William Harris Hardy, the road's promoter, stated that the project was conceived on the proposition that timber was becoming increasingly valuable and that a demand would soon arise for the products of the longleaf forests. He correctly envisioned the development of a great forest industry in the vacant pine woods and a profitable railroad operation arising from it.[2]

In the years 1896-1903 the Gulf and Ship Island Railroad was built from Gulfport on the Mississippi Coast to Jackson, and it became one of the most important lumber carriers in the nation. The acquisition of the logging road of the Eastman-Gardiner Lumber Company gave the Gulf and Ship Island a branch line to the important lumber production center of Laurel. Another logging road constructed by the Camp and Hinton Lumber Company came to be one section of a branch line of the Gulf and Ship Island extending from Maxie to Columbia and thence northward to Mendenhall, where it rejoined the main line. With its branches the Gulf and Ship Island had a total mileage of 306.67 in the virgin forest country lying south of Jackson to the Gulf Coast, between the Pearl and Pascagoula rivers. From Ship Island to the mainland the company dug a channel, so that at the Gulfport harbor lumber was loaded directly from railroad cars into cargo ships.[3]

Another important lumber carrier, the Mobile, Jackson,

and Kansas City Railroad was projected in 1870 chiefly to exploit the pine forests lying northwest of Mobile in Alabama and Mississippi. Construction got under way but was soon abandoned. Not until the demand for yellow pine became greater in the eighties was the railroad project revived, but again little was accomplished. In 1896, after Col. Frank B. Merrill assumed its direction, the road was built to Laurel and then due north to Jackson, Tennessee, passing all the way through the pine woods of eastern Mississippi.

The New Orleans Great Northern was built and financed chiefly by F. H. and C. W. Goodyear and other lumbermen. Early plans of the company called for construction of a railroad from New Orleans to Bogalusa and thence up the west side of the Pearl River to Jackson, Mississippi. Such a road would run for most of its length through a virgin forest lying between the old Illinois Central and the Pearl River. In 1909 the road was completed; in the next year 77 per cent of the road's total tonnage consisted of timber products.[4]

Along the main line railroads large sawmills were erected. Then, after the timber near the mills was exhausted, logging roads often more than thirty miles in length were extended out into the virgin forest. At first they served no other purpose than to transport logs to the mill. But as the roads grew in length, small sawmill towns and villages often sprang up near them, creating a demand for transportation service of a general nature. Timber cutting created vast areas of cutover land, some of which was converted into farms. Thus the logging roads provided means for populating the bare country, and some of the lines because of the nature of their traffic became common carriers. Although some of these were incorporated and came to offer general transportation services to the areas in which they were located, their principal traffic in every case continued to be the logs of the lumber companies that owned the roads.

The Natchez, Columbia, and Mobile Railroad belonging to the Butterfield Lumber Company of Norfield and incorporated in 1892 was one of the first logging roads converted into a common carrier. Leaving the Illinois Central at Norfield, the road extended eastward fourteen miles to a point called Main Junction, where it divided into two branches, one extending northwestward for nine miles, the other going seven and one-half miles to Furlough Switch.

Another road, the Mississippi Central, owned by the J. J. Newman Lumber Company whose large mills were located at Hattiesburg and nearby Sumrall, ran from Beaumont, where it connected with the Mobile, Jackson, and Kansas City Railroad, to Natchez. This road, which grew out of the old logging road of the Newman Lumber Company, opened up virgin longleaf and shortleaf areas between Hattiesburg and Natchez. In 1906, before it was completed, there were twenty-six sawmills on its lines; in 1913 there were twenty-five. The road gave these mills outlets to Gulf ports and to the areas beyond the Mississippi River.

Up to 1880 the J. J. White Lumber Company of McComb operated a narrow gauge railroad westward from McComb to its timber holdings. The people of Liberty, the county seat of Amite County, induced the lumber company to extend its road into the town. When incorporated in 1902 the road switched from narrow to standard gauge rails. Later in 1907 the road was built east of McComb nine miles to New Holmesville; a branch line was extended south fifteen miles from Irene, a point between Liberty and McComb, to Keith. Although the road provided general transportation services, its chief function continued to be one of supplying logs to the J. J. White mill.[5]

The Fernwood and Gulf Railroad, together with the Fernwood Lumber Company, was owned by the Enochs brothers. This road had begun as a logging road using wooden rails. In 1906 it was incorporated, but the Fernwood Lumber Company reserved the right to transport its logs over the line without cost. The thirty-three miles of road operated by the company ran from the company mill at Fernwood on the Illinois Central Railroad to Tylertown, where it connected with the New Orleans Great Northern Railroad, and thence northeastward one mile to Kokomo.[6]

A number of other small incorporated roads were built by lumber companies in Mississippi. The Mississippi-Eastern Railroad, 16.2 miles in length, owned by the Mississippi Lumber Company and originally a logging road, was projected to run to a point near the Tombigbee River. At Chicora on the Mobile and Ohio the Robinson Land and Lumber Company built a short road known as the Chickasawhay and Jackson Railroad. The W. Denny Lumber Company logging road, which became the Pascagoula and Northern Railroad, connected Moss Point and Lucedale. Following the failure of the W. Denny Company, the railroad was acquired by the Dantzlers.

Lumbermen and trunk line railroad owners, despite their community of interest and interdependence, were often involved in heated controversies. One of the lumbermen's chief grievances against railroads in the years 1900-1915 was freight rates. Other problems of transportation were of long standing. Attempts by both the Federal and state governments had been of little avail in establishing such regulation of the railroads as would serve the interest of the shipper. The Interstate Commerce Act of 1887 and the Elkins Act of 1903 attempted to control the worst abuses by the railroads. But the Supreme Court through a number of decisions left the railroads largely unregulated. In spite of the provision of the Elkins Act against rebates and rate discriminations, deviations from published rates continued to be allowed by some of the railroads in the South both east and west of the Mississippi River.[7]

Yellow pine producers in one section of the South competed against manufacturers of a similar product in other areas. In this situation the policies of the various railroads could give a competitive advantage to a favored southern pine area with respect to northern markets. Lumbermen with access to lower freight rates might easily undersell their competitors. In 1894 several millmen on the New Orleans and Northeastern who claimed to represent an investment of $2,000,000, most of whose lumber was shipped north of the Ohio River, requested a freight rate reduction. Other groups of lumbermen, suffering from what they considered unjust rates, appointed I. C. Enochs and George S. Gardiner to confer with railroad officials with the object of obtaining lower rates.

Apparently direct negotiation with railroad officials was generally unsatisfactory, for in 1902 a resolution approved by the Southern Lumbermen's Association requested enlargement of the powers of the Interstate Commerce Commission. According to the committee on resolutions, past efforts of the Commission to end discrimination and other evils had failed. Railroads had done little to settle grievances. On the one hand, roads pleaded inability to change matters on the ground that state action tied their hands. On the other, state railroad commissioners contended that their authority did not extend to interstate commerce. Nor up to this time had the Interstate Commerce Commission been effective. Of the sixteen principal cases appealed to the Supreme Court in the years 1897-1906, all but one had been lost by the Commission.[8]

In 1903 a long struggle over freight rates and the so-called tap-line divisions got underway. These were closely interrelated, since both had a direct bearing on competition between yellow pine lumbermen west of the Mississippi with those east of the river in the sale of lumber in the Ohio River region. This important consumer section was known as the Central Freight Association District, bounded on the south by the Ohio River, on the east by a line running through Buffalo and Pittsburgh, and on the west by the Mississippi River. In this area large quantities of lumber produced in the interior of Mississippi were marketed.

Freight rates fluctuated in the years prior to 1894. Yellow pine producers west of the river secured an important advantage over their competitors on the east side by obtaining a two-cent differential on each hundred pounds of lumber. To put the two regional groups of competitors on an even basis, the freight rate to Cairo, Illinois, a basing point, was made uniform for millmen on both sides of the Mississippi. At Cairo, lumber traffic from both sides of the river converged en route to the Central Association District, In so far as the published rate to Cairo was concerned, rates were the same; but in reality, because tap-line division (a subject to be discussed below) was extended to many lumbermen west of the river but withheld from others and from most millmen on the eastern side, the tariff remained unequal.

During the years 1899-1903 lumbermen enjoyed prosperity. An increase in the freight rate to Cairo from thirteen to fourteen cents in 1899 met no organized or concerted opposition. This prosperity, however, may have motivated the railroads suddenly to increase the fourteen-cent rate to sixteen in April, 1903. The higher tariff went into effect on both sides of the river on roads comprising the Southeast Mississippi Valley Railroad Association. Lumbermen east of the river, through their representatives Gardiner and Enochs, met informally with members of the Interstate Commerce Commission and officials of railroads with a view to settling the grievance amicably. The efforts of Gardiner and Enochs were fruitless.[9]

Many of the large operators east of the river got together and prepared for a showdown fight. They organized the Central Yellow Pine Association to represent a membership possessing capital of $50,000,000 and an annual productive capacity of 1,500,000,000 board feet valued at $7,500,000. It was estimated that the two-cent freight hike on 1,500,000,000 board feet shipped from Mississippi and east Louisiana to the interior would increase transportation costs by about $1,000,000.

The first move in the fight against the railroads was a request to Judge H. C. Niles of the Southern District Federal Court in Mississippi to enjoin the carriers from collecting the advanced rate. Roads comprising the Southeast Mississippi Valley Railroad Association were accused of having formed an illegal combination in restraint of trade and of having violated the Federal antitrust statutes by agreeing on a uniform rate which was not the result of competition. Judge Niles ruled that his court lacked jurisdiction. To issue an injunction against the increased rate would in effect be the same as fixing the tariff. Authority to fix rates was a legislative function not resting in the court nor within the purview of the Interstate Commerce Commission. Only Congress possessed the authority to determine rates.

Lumbermen took the next step in the controversy by

placing their case before the Interstate Commerce Commission. The tap-line divisions, which were extended to some lumbermen but withheld from others, were a feature of the problem upon which the Commission was asked to pass. Testimony was heard by the Commission in 1904 and early 1905. The bill of complaint submitted by the lumbermen contended that the freight rate on yellow pine was unreasonable, injurious to them, and in violation of the Interstate Commerce Act. Markets formerly accessible were contracted; equal competition in the Central Freight Association District had been destroyed. Excessive rates had been imposed upon yellow pine more grievous than those on other products of like weight and tonnage. Even though the rates were nominally the same for producers on both sides of the Mississippi River, for those who received tap-line divisions the rates were lower.

The decision of the Commission was a victory for the lumbermen. According to its findings the rate advance was not the result of competition. The fourteen-cent rate was reasonable. Test of the reasonableness of a particular rate lay not in the profit of the shipper but in its yield of a reasonable return to the carrier. The Commission found the rates on lumber in effect prior to 1903 reasonably high in comparison to other commodities which were relatively analogous to lumber as to volume and other conditions affecting transportation service. No rule, according to the Commission, was more firmly grounded in reason or more universally recognized by carriers than that the greater the tonnage of an article of traffic the lower the rate. Yet yellow pine had been made an exception to the rule. In the opinion of the Commission a lumber tariff should be comparatively low because lumber was not perishable, required no special equipment, furnished a large tonnage, was shipped year in and year out at all seasons, and was loaded by shipper and unloaded by consignees. The two-cent rate increase of 1903 was unjust and unreasonable.

When roads were directed by the Commission to cease collecting the increased rate, they refused to obey the order. To enforce its ruling the Commission asked the Federal Court of Appeals for an injunction against the roads. Judge Charles Parlange ruled against the roads upholding the decision of the Commission. The roads then appealed to the United States Supreme Court.

Lumbermen were dissatisfied with the legal procedure involved in dealing with the railroads. The rate controversy and other grievances caused lumbermen to favor giving the Interstate Commerce Commission real powers in regulating the railroads. A resolution on this subject by the Southern Lumber Manufacturers' Association in 1902 has already been noted. Representative attitudes of yellow pine lumbermen were again expressed in a memorial to Congress in 1904. In substance the lumbermen asserted that the existing system of regulation was imperfect, for neither the Commission nor the courts could protect the people against unlawful exactions and discriminations in rates. The people were powerless to protect themselves under existing remedies or to cope with the powerful combinations among railroad companies which had in the past three years unlawfully filched $300,000,000 from the public. In the opinion of the lumbermen the only agency having the knowledge, integrity, and willingness to deal with the roads was the Interstate Commerce Commission. The lumbermen asked that the Commission be given the authority to fix a reasonable rate that would stand until struck down by the courts.

Later in 1907 lumbermen again advocated that enlarged

powers be granted the Commission. R. A. Long contended that already too much legislation existed; I. C. Enochs made a heated reply to his objections, accusing the roads of failing to prepare for the growth of industry and of using their huge profits to purchase other railroads rather than to provide better transportation service for the people. As for himself, stated Enochs, he feared neither additional legislation nor politicians. Liberty as people knew it had resulted from wise legislation; there had been too few rather than too many laws. Lumbermen supported the position of Enochs by a vote of thirty-five to twelve.

A few months before the rate controversy came before the Supreme Court, the so-called Hepburn Act amending the original Interstate Commerce Act of 1887 was enacted by Congress. Previous acts to give the Commission greater regulatory authority had been largely nullified by numerous adverse Federal court decisions. The courts had exercised the right to view Commission decisions both as to fact and law. They had admitted new evidence in appellate cases which in effect virtually amounted to new trials. The ineffectiveness of the Commission is illustrated by the fact that thirty-two of its decisions had been reversed on appeals to Federal courts. In twenty-six of these cases the courts had taken additional evidence and had refused to accept findings of fact by the Commission as prima facie evidence. The Hepburn Act of 1906 gave the Commission ultimate but not original rate-fixing powers.[10] The Commission could take any rate under review on complaint, decide a fair and reasonable maximum, and order the railroad not to charge in excess of it.

After much debate the power of the Federal courts to exercise broad review over the decisions of the Interstate Commerce Commission had been left unchanged by the

Hepburn Act. Yet shortly after passage of the act the Supreme Court in a series of decisions voluntarily curtailed its power of review. The first indication of a change of attitude came in the lumber case *Illinois Central Railroad vs. Interstate Commerce Commission,* 1907. The Court held that it would not investigate and review all the facts of a case on appeal. I. M. Miller stated to the Court that the reasonableness of a rate was a question of fact which the Commission was peculiarly fitted to determine. The Court held that the Commission had probative force and its finding of fact was by law prima facie evidence. The Court would not review such findings when they were concurred in by a Federal circuit court of appeals unless the record established clearly and unmistakably that error had been committed. The Court concurred in the decision of the Commission that the two-cent rate hike was unwarranted and unreasonable. It was a victory for both the lumbermen and the Interstate Commerce Commission. In 1910 the Court again in a more elaborate decision approved the principle of narrow review embodied in the Long amendment.

Mississippi lumbermen hailed the Court decision as a victory in their long struggle with the railroads over rates. Many millmen believed that the authority of the Commission ought to be further expanded to include the power to fix maximum and minimum rates and to forestall rate increases going into effect prior to approval by the Commission. In the Mann-Elkins Act of 1910 the Commission gained the authority to suspend proposed advances in rates pending investigation of their propriety.

Closely interwoven with the rate problem was the practice by main trunk-line railroads west of the Mississippi River of granting to millmen adjacent to their lines a division of the freight revenue for transportation of logs to mills. For producers who received such an allowance from the carriers, freight costs on lumber were in effect reduced by the amount of the division, as millmen east of the river saw it. Although practices varied, lumber was usually billed from the junction point of the tap line and the main line, where the mill was almost always located, and the full rate was paid on the shipment. The trunk-line road compensated the lumber company periodically by returning a portion of what had been part of the cost. Though the division of the through rate returned to millmen was unlikely to be as great as the actual expense of transporting logs to the mill, it amounted in some instances to between two and seven cents on each hundred pounds of lumber shipped. This practice was defended as lawful under the milling-in-transit principle. Generally this principle applied to a situation in which raw materials paid the local rate to the point of manufacture. Afterward the finished product was carried upon a tariff calculated as if the product had originated at the point where the raw materials first entered transportation channels.[11]

The practice of granting tap-line divisions had developed gradually during the years 1888-1894. Lands had been bought and mills erected upon the understanding that a division of the freight revenue would be granted. After the uniform freight rate to Cairo was established, railroads west of the river extended divisions to many millmen of that section. Gardiner of the Eastman-Gardiner Lumber Company then called upon the New Orleans and Northeastern Railroad to accord him the advantage enjoyed by his competitors on the west side of the river. His request was approved, and for a brief period millmen east of the river received divisions. However in 1903 only those lumbermen adjacent to the Mobile and Ohio were compensated by the trunk line.

In 1903 Mississippi millmen asked the Commission to declare tap-line divisions illegal or to make them general practice. Divisions were, they argued, basically rebates and as such were prohibited by the Elkins Act of 1903. While published rates and manufacturing conditions were the same for all millmen, producers who received tap-line divisions actually enjoyed lower freight costs, a fact which promoted unfair competition in the Central Freight Association District and elsewhere.[12]

The defendants denied that tap-line divisions were rebates in violation of the law. They contended that tap lines performed the same services as common carriers; at the time logs were loaded in the woods they entered interstate commerce. According to this view, the lumber originated not at the mill but in the forest; therefore, divisions were legal.

With some important reservations, the Commission in 1904 approved of tap-line divisions. Transportation of a log to the mill by one line and the carriage from the mill by another road could be treated as a through shipment of lumber. Allowance for moving the logs to the mill would be the same as if the lumber originated at the point where the logs were received, provided both roads were common carriers by rail. Common carriers must submit statistical reports and file tariffs according to law, and be subject to public control. The Commission acknowledged that treating logs as lumber for the purpose of establishing a joint through rate involved an extreme extension of the milling-in-transit principle.

The 1904 tap-line decision was reviewed briefly by the Commission in the freight rate ruling of 1905. Tap-line divisions extended by the Mobile and Ohio road were declared illegal on the grounds that the logroads receiving compensation performed no other service than that of transportation of logs to the mills and were the private property of lumbermen.

In light of the decisions of 1904 and 1905, millmen on both sides of the river proceeded to transform their logroads into common carriers. One of the major steps in this process was the incorporation of logroads and the separation of ownership of the roads from that of the sawmills. Such action was a legal fiction designed to enable the millmen to participate in rate divisions. In most instances the great bulk of traffic moving over the tap lines continued to be the logs of the mill company.

In 1908 and 1909 the tap-line question was again up for consideration before the Commission. The Commission recognized that millmen had incorporated their logroads to obtain divisions, though the character of traffic and ownership of the tap lines remained unchanged. According to the Commission few of the so-called tap lines met the test of a common carrier. Of the 698 tap lines considered, lumber furnished the entire traffic on 498; 80 per cent or more of the traffic on the remaining 200 lines derived from the lumber interest that owned or controlled the line. As many as 621 of the tap lines neither filed tariffs of their own nor concurred in tariffs of other lines concerning interstate shipments. Only 92 of the tap lines filed reports with the Commission. Only 183 had undertaken the formality of being incorporated as railways.

The Commission held that it would not recognize as common carriers lines which failed to publish tariffs in lawful form or to concur in joint rates of other lines. Moreover, lines failing to file annual reports and to keep their accounts in accordance with rules laid down by the Commission would be denied classification as common carriers. Even tap lines which met these requirements could not be considered common carriers if they had no traffic other than the logs of the company which owned both mill and log-road.

Only a small number of tap-line owners in Mississippi were affected by the decision, but indirectly all millmen had a stake in the controversy. For those lumbermen who were recipients of divisions the cost of production was less, and they were consequently in the position to undersell competitors in northern markets. Some millmen believed that the tap-line divisions enabled lumbermen west of the Mississippi to operate their mills day and night, thus over-supplying the market and at times dumping large quantities of lumber in that section of the country where most of the lumber produced east of the river was sold.

On protest of lumbermen the Interstate Commerce Commission delayed cancellation of divisions until 1911. In that year the Commission again reconsidered the question in its relation to yellow pine manufacturers in Louisiana, Arkansas, Missouri, and Texas. The resulting decision of the Commission in 1912 undertook to lay down principles for determining whether a tap line was a common carrier or an industrial or plant facility. On the one hand, ruled the Commission, the fact that a mill company owned a road was of itself insufficient grounds to divest the line of its character as a common carrier. On the other hand, if a line had been turned over to an incorporated company but still operated in the interest of the lumber company, it retained the status of a plant facility. Any service performed by a line for the mill that controlled it could not be regarded as transportation, and any division made to such a line was a rebate. It was held not permissible for a trunk line to honor a milling-in-transit privilege by which the lumber rate was extended over tap lines out into the forests where the logs were cut, unless the trunk line pursued the same course with respect to logs cut along its own line. The effect of the decision was to restrict narrowly the milling-in-transit principle and to abolish divisions extended to tap lines whose main traffic was lumber owned by a particular lumber company.

The Illinois Central had been allowing the Natchez, Columbia, and Mobile tap line two cents per hundred pounds for traffic originating on its line. This tap line was owned by the Butterfield Lumber Company, which also provided the bulk of its freight. In effect the lumber company received not only two cents per hundred pounds of logs hauled to its mill, but also another two cents on the lumber which independent mills shipped over the tap line that was charged at the junction point of the Butterfield Railroad and the Illinois Central. The Commission ruled that no compensation could be allowed the tap line on products owned by the Butterfield Lumber Company. A similar ruling was applied to the Liberty-White and the Fernwood and Gulf railroads.[13]

The tap-line owners refused to accept the decision of the Commission as final and in 1914 appealed the case from the Commerce Court to the United States Supreme Court. Before the Court was placed the question whether tap lines were common carriers or plant facilities. The Commission had found that most of the traffic on the tap lines consisted of logs and other properties of the proprietary mills and that only a small portion was owned by others. The Court held that the real criterion was not the amount of non-proprietary traffic carried by the road, but rather whether the public could demand and receive its service. The Commission could, however, act to reduce the amount of divisions, control abuses, and end discrimination.[14]

A third important grievance held against railroads by lumbermen arose from the shortage of freight cars. This problem was of long standing, dating back to the early nineties or before. Lack of cars became extremely acute at times after the turn of the century, when lumber production assumed immense proportions. Lumbermen in 1901 claimed that they were the last to obtain cars and accused the roads of favoritism. In the next year the shortage was attributed to the blindness of railroad officials who had failed to foresee and prepare for the tremendous industrial development the country was then experiencing. The weight capacity of cars, it was true, had been increased. But the increase did not enlarge their storage capacity, and it therefore benefited lumbermen little. Indeed, the greater weight of loaded cars retarded transit and thus made the acute car shortage still worse.

During the fall and winter of 1902 millmen claimed that they obtained only half of the transport capacity they actually needed. Lumbermen in 1903 had to decline some orders because they could not promise quick shipment. According to one writer, cars were not delivered on request and requisition, congested conditions delayed rail movements, and the transportation problem was making southern pine a poor competitor. A further difficulty was the refusal of some of the railroads to allow their rolling stock to move on the lines of other roads.

In the latter part of 1906 the shortage of transportation facilities was reported to be without parallel, with only one-third of the car requirements of millmen being met. Although orders were plentiful and demand high, because of the extreme car famine the cut exceeded shipment. On September 25, 1906, the Illinois Central could offer lumber shippers but nine cars, and they could be routed only to points touched by the railroad. Some of the small mills were unable to operate continuously; those cutting less than 25,000 board feet daily were generally compelled to sell their products as they came from the saw.

Little if any improvement in the car situation was discernible in the early part of 1907. Millmen complained of receiving only 20 to 30 per cent of normal car requirements. In the following fall and early winter the perennial seasonal shortage was exceptionally severe. Some millmen attributed the crisis to inefficient railroad management. C. W. Robinson asserted that loaded cars moved on the average about twenty-three miles per day.

According to I. C. Enochs, it required 200,000 cars annually to move the three billion board feet of yellow pine lumber. The business of a lumberman was contingent upon an adequate car supply. While the dividends of roads had increased because of higher tonnage and rates, the quality of service had declined. Huge profits made by railroad corporations ought to have been invested in better service rather than in acquiring stocks of other corporations.

Scarcity of rolling stock was generally more pronounced on short-line railroads. The Illinois Central and the Mobile and Ohio roads, with terminal points in northern cities, could control the distribution of cars on such lines as the New Orleans Great Northern, the Gulf and Ship Island, and New Orleans and Northeastern. Outgoing lumber shipments exceeded by far the total incoming traffic, an imbalance which contributed much to the shortage of transportation facilities on the short lines.

Much lumber was shipped on flat cars or in open gondolas. Stakes and binders furnished by the lumbermen at their own expense were used to secure the cargo on the open cars. Annual cost to the lumber industry for equipping flat cars, according to Robinson, ranged from $50,000 to $100,000 or from $2.50 to $3.00 per car. Besides this, the weight of the stakes and binders for a time formed a part of the freight charges. The cost of stakes plus freight charges on them was computed at $5.94 per car in 1904.

Lumbermen sought by negotiation with the roads and finally by an appeal to the Interstate Commerce Commission to induce the carriers to assume the cost of staking cars. The railroads responded by granting a five-hundred-pound weight allowance for stakes and binders but refused to bear any other expense. The Commission held that no accurate method of determining the cost of staking existed. One of the chief difficulties was that no permanent stakes had been invented. It had long been customary for the shipper to furnish the equipment to secure lumber on flat and gondola cars. Lumber, logs, and cordwood were loaded differently; it was considered impossible to devise permanent stakes which would secure all types of forest cargo. Again in 1915 lumbermen sought to compel the roads to bear the cost of equipping flat cars. The railroads, though opposing the move, expected to lose their case because permanent stakes by this time had been devised. But the Commission refused to set aside its decision of 1908.

Development of the lumber industry provided the principal motivation for railroad building in the pine country. By 1910 the transportation system had penetrated almost every locality. The services provided by the railroads were, however, generally unsatisfactory to interior lumbermen, for in the interior the roads possessed a transportation monopoly; they were for the most part indifferent to the lumberman's grievances and at times wholly ignored them. The lumbermen were thus led to seek assistance from the state and Federal governments. Although disposed to be strongly individualistic and *laissez-faire* in their economic outlook, they used every means possible to bring about Federal regulation of the railroads. Through litigation in the courts over rates and by other activities the millmen had a significant influence both directly and indirectly in bringing about an enlargement of the powers of the Interstate Commerce Commission.

CHAPTER XVI

THE LABOR PROBLEM

T HE DEVELOPMENT OF FOREST INDUSTRIES in Mississippi brought social and economic changes to a section of the South that had advanced little beyond the pioneer stage. The native white herdsmen were compelled by the decline of their age-old economy to become, often unwillingly, a part of the industrial system. Like the scattered small farmers who were also drawn upon by the lumber industry, they had been bred in remoteness and isolation, and both herdsmen and farmers were ill-conditioned to live effectively in the regimented world created by modern industry.

In the early years after the Civil War, while the lumber industry was still relatively small and undeveloped, the local population met all requirements for labor. There was available for employment in the coast mills a considerable number of freedmen who had become mill hands in the ante-bellum period. Poitevent and Favre, William Griffin, the Westons, and possibly the W. Denny Company employed a number of Negro mill hands whom they had trained in the ante-bellum period. The knowledge and skills of these early sawmill workers were passed down to others; hence the coast mills were, unlike those established in the interior, seldom handicapped by labor shortages.

As the industrial system gained momentum on the Gulf Coast, it created many problems unknown to the small mill operator in the ante-bellum period. Then the average mill-man had either owned his own labor or known his hired employees by their first names. But around the turn of the century and even earlier on the coast, large manufacturing establishments which required a considerable number of laborers created a psychological barrier between workers and owners. As the manufacturing plant increased in size and production became more specialized and complex, the owner had less time to consider the individual needs of the worker. No longer could employers know employees by name. The division of labor in the mills and delegation of authority to subordinates made each laborer a hireling of the department head. It became impossible for the slab hustler to present his grievances, if he had any, directly to the owner for consideration.

Despite the impersonal human relationships characteristic of the industrial system, a few of the coast lumbermen continued the paternalistic policies of the petty capitalist in respect to their employees. Henry Weston, a self-made capitalist, and his seven sons associated with him in the mill business were especially sensitive to the welfare of their employees. Similarly the L. N. Dantzler Company, which came to own several mills and employ a large number of workers, remained paternalistic in its policies. Difficulties between foreman and worker were often settled by the general manager or others high in authority.[1]

Since most lumber products were generally low in price

from 1865 to 1890 and timber was cheap, labor constituted one of the most considerable costs of production. Millmen thought that wages must of necessity be low and hours of work long. The wage scale for mill workers varied from locality to locality depending upon the supply of labor and the demand for it. In 1880 the average wage for a twelve- to fourteen-hour day in the coast mills and those in Pike County was $1. Among the unskilled, the average was probably much lower than that reported in the census figures. In 1889 common labor was paid $20 per month and skilled workers about $2.50 per day. Occasionally millmen short of cash were unable to pay off their workers. When this happened, all that the worker received for his labor was food and clothing. Even after 1889 a wage of fifty to eighty-five cents per day for common labor was not unusual.

During the seventies the working day was often longer than fourteen hours. Men entered the mill before break of day to remain at work until after dark. The employer furnished his workers breakfast and lunch. It was said of the L. N. Dantzler Lumber Company that it had no difficulty in obtaining and keeping workers, since it was the only company which supplied meat with meals. Extremely long hours of work and irregular paydays producing only the bare necessities for a worker and his family were general in the coast mills up to 1900.

In the late eighties the coast sawmill workers undertook to improve their working conditions through organized effort. Through their union, the Knights of Labor, they demanded that the workday be shortened to twelve hours. Failing to achieve their objective by peaceful negotiations, the workers in the Moss Point-Pascagoula area went out on strike. After a few days of work stoppage the millmen capitulated to union demands and the strike ended. Flushed by their victory, the Knights of Labor proceeded to organize large numbers of the mill workers.

Two years later the Knights of Labor boldly demanded reduction of the workday to ten hours, payment of wages once each week, and, finally, recognition of the union as a bargaining agent. All three points were rejected, and three hundred mill workers struck on February 22, 1889. Public opinion as interpreted by the Pascagoula *Democrat Star* was aligned on the side of the mill owners. The editor predicted that the strike would fail because starvation would compel the workers to return to their jobs. Even when working, according to the editor, the laborers were able to obtain only the bare necessities.

Intelligent union leaders were accused of inciting the strikers to violence by playing upon their ignorance and fear. John Broadean, the Negro president of the colored division of the Knights of Labor in the Moss Point-Pascagoula district, was reported to have urged the Negroes to violent action by saying: "The Knights of Labor live together, work together, and if necessary, can die together."[2]

The strikers lacked the resources to maintain an extended strike. The union tried to fill in the gap caused by the loss of wages by issuing a weekly ration of five pounds of meat and a peck of corn meal to each worker's family. Early in March only seventy-five of six hundred men were working in the mills; strikers were receiving provisions, but no money, from outside sources.

To end the strike workers were imported. The sheriff deputized thirty-one men to assist him in protecting them. The strikers threatened those who returned to work, but apparently without effect. In mid-March all the mills were operating with a full complement of labor.

Unrest was general among sawmill workers along the Gulf Coast. At Handsboro in 1888 the mill hands led by the Knights of Labor struck for shorter hours of work and recognition of their union. Liddle and Hand came to terms with the union, but Henry Leinhard refused to negotiate. For a short period Leinhard's mill was idle, but when he agreed to a ten-hour workday the strike ended.[3]

The next few years were peaceful. If the laborers were dissatisfied with working conditions, they lacked both leaders and a strong organization to make an effective protest. The twelve-hour workday and infrequent paydays caused by the inability of millmen to compensate their workers in cash continued in the Moss Point-Pascagoula district. The Knights of Labor, relatively inactive for ten years, in 1899 felt strong enough to attempt to compel the millmen to reduce the workday and institute weekly payment of wages. In addition the union again demanded recognition as a bargaining agent. Refusal of the millmen to accept the terms brought on a strike. Mill owners, long aware of a potential strike, had accepted orders from their customers on the condition that, if a work stoppage occurred, no lumber would be delivered.

About five hundred workers, mostly Negroes employed in the Denny, Dantzler, and Bounds mills, quit work in April, 1900. Union leaders put forth considerable effort to convince the public of the justice of their cause and of their peaceful intentions. But all did not go as planned. In May, Negroes in considerable numbers began to resume work. White workers were accused of intimidation to prevent the blacks from returning to the mills. Those who remained idle were threatened with a loss of their jobs by the mill owners. To break the back of the strike, the millmen obtained a court injunction against the union, prohibiting it from using violence to prevent workers from returning to their jobs. When some of the union members started working, they were fired upon by alleged union agents.

Public opinion, never particularly favorable to the strikers, became openly antilabor. Mark A. Dees, a businessman-legislator, asserted that he had supported peaceful efforts of the sawmill laborers to improve their lot, but he could not approve of violence. At Moss Point an extralegal body of citizens known as the Committee of Public Safety was formed to take drastic action against the labor leaders. Within a few days about a dozen people suspected of being dangerous characters were ordered to leave town, and four men were arrested and accused of attempting to assassinate mill employees. Eight union members convicted of firing upon workers were each given ten years in the penitentiary. So ended not only the strike but also the attempt to organize coast mill workers into a labor union. Henceforth, improvement of working conditions was to come by legislation and through competition between millmen for labor.[4]

The failure of the labor movement in the coast mills probably governed what happened in the interior, where the lumber industry was in its infancy in the decade 1890-1900. Interior mill workers made little if any attempt to organize in the years 1900-1915. Obstacles to creating a class-conscious group of workers were too many and too great to be overcome. Negroes composed a majority of the mill workers in the interior. Native whites were next in number. Northern whites, mostly skilled workers, and a sprinkling of foreigners constituted the remainder. The native white, strongly individualistic, was poor material for the discipline and co-operation demanded of a successful labor union. Moreover, the presence of the Negro as a competitor for the unskilled and frequently for the skilled positions created a division of labor into two ranks often bitterly hostile to each other. In general the mill owner was compelled to reserve to the white man all the better-paid jobs and those that included supervision, leaving to the Negro the low-paid tasks and those that were extremely hazardous to life and limb.

Competition between white and Negro was one of the factors that gave rise to the White Cap movement in southern Mississippi during the 1890's. The high tide of the White Cap movement came in the years 1891-1894. The elements that contributed to the White Cap movement were substantially the same as those which later constituted the Ku Klux Klan in the 1920's. The group vented its spleen against merchants, industrialists, foreigners, Jews, and minority groups. The White Cappers declared in their constitution a determination to keep the Negro on the farm by persuasion, and if necessary by coercion.[5]

On December 10, 1893, the engine house of the Norwood-Butterfield Lumber Company at Brookhaven was burned by the White Cappers. According to the company, other heavy losses had been incurred earlier at the hands of the organization. Because of the lawlessness of this irresponsible group, the company had decided that investment of capital was risky in Mississippi. In northern Jackson County, where Gregory Luce employed a number of foreign workers in his logging crew, native whites calling themselves White Cappers posted notices in the woods warning the newcomers to flee or take the consequences of staying and being subjected to powder and lead. The warning led to the posting of guards in the woods to protect the foreigners.[6]

The recruitment of a large labor force was necessary to the development of the lumber industry in Mississippi. In 1899 there were 16,421 sawmill workers in Mississippi, mostly in the longleaf pine country; in 1904, 24,415; in 1909, 37,178.[7] This mushrooming demand in a sparsely populated region where few of the inhabitants possessed the skills to operate the machines of the modern sawmill produced an acute labor shortage. As manufacturers moved their plants into virgin pine country, they were compelled to recruit a considerable number of workers from other parts of the country. Thus the older lumber regions in the North and East provided not only most of the initial capital and managerial skills for the industry in Mississippi, but also most of the key workers. Usually when the northern lumberman moved south, he brought such men with him to operate his mill. I. C. Enochs stated that wherever sawmills had been erected, it had usually been necessary to bring in department heads and skilled workers from other sections of the country. When the Finkbine Lumber Company erected a large mill at Wiggins in 1902, almost all the department heads and skilled laborers were recruited from the North. Generally the local population, both colored and white, provided the bulk of the unskilled common labor. In 1895 a majority of those engaged in manufacturing, logging, and rafting in the South were said to be Negroes. After a sawmill had been in operation for a few years in a locality, the key men were increasingly recruited from the local population. The southern workers, both colored and white, were mechanically inclined and needed only an opportunity to develop their latent capabilities.

For most of the years 1898-1907 there was an acute shortage of both common and skilled workers. Moreover, according to millmen, the available supply was both inefficient and undependable. In 1900 S. C. Eaton, a lumberman near Hattiesburg, reported a need for two hundred additional workmen. Many of his Negro workmen were leaving the woods and mills, and because of their instability, he had decided to replace them with white men. One writer stated in 1901 that unskilled labor in the woods and sawmills was seriously inadequate and that there was no known remedy for the shortage. In the sawmills and logging camps adjacent to the Gulf and Ship Island Railroad, workers were scarce in October and December, 1900. Most of the workers were Negroes, and the higher the wages paid them, the more unreliable and unsteady they became. A few who operated machinery were well paid and remained on their jobs; but the problem arose with the common laborers who, when paid for two or three days, quit for the rest of the week.

The editor of the *Lumber Trade Journal* reported that some lumbermen in 1904 had an adequate supply of labor, but others were short. Every one acquainted with manufacturing knew, he said, that the Negro was not entirely satisfactory as a worker. In October, 1904, J. J. White of McComb asserted that labor was short and the available supply not very good. The Negroes would work only half of the time and were difficult to control. With increased wages it was necessary for them to work only half time to earn a bare living.

One observer contended that the high price of cotton was a factor in promoting instability among Negroes. Delta planters competed with the millmen for Negro labor, and the black man would choose cotton picking every time in preference to sawmill work. On the cotton plantation rations were furnished, and with very little labor the Negro could make a crop. I. C. Enochs also asserted that the labor shortage arose from the high price of cotton and competition between millmen and planters. He thought that the Negro usually followed the line of least resistance and would prefer to grow cotton.

Labor shortages and the difficulty experienced in keeping the Negro on the job led to the enactment of a drastic vagrancy act by the state legislature in 1904. The act, a revision of earlier legislation on the subject, gave a broad definition of vagrancy. The clause, designed to compel the Negro to labor continuously, classed as vagrants persons without visible means of support and only occasionally employed. Vagrants were to be convicted and imprisoned for not less than ten nor more than thirty days. Money payment in lieu of jail sentence was prohibited. The penalty for second offenders was ninety days to six months in jail and payment of all court costs. The Vagrancy Act contributed nothing to the alleviation of the labor scarcity. Quite the contrary, the act caused hundreds to leave the state, making an already bad situation worse.

The worker shortage was such that many mills in 1904 were scarcely able to conduct full scale operations. Lack of labor was blamed for the closing of the Erata Lumber Company. After a work stoppage at the Kingston Lumber Company at Laurel, laborers were given weekly paydays.

A writer in the fall of 1905 reported a familiar observation: industry was booming, but lumbermen were hampered by an acute shortage of labor; Negro labor was scarce and undependable. According to J. J. White, Negroes would work only about one-third of the time; many of them would not labor at all. He had worked Negroes both when they were slaves and after they were freed. As a workman the Negro had been getting worse every year since he had obtained freedom. White asserted that labor, both white and colored, in his area was scarce, shiftless, and unreliable. An official of the Camp and Hinton Lumber Company of Lumberton complained that the firm had employed as many as six hundred Negroes at one time and that they were becoming "more no-account and trifling every day." W. E. Guild, Treasurer of the Finkbine Lumber Company, a northern concern, and part owner of the large mill at Wiggins, commented on the fact that the Negro was "unstable and unreliable as a laborer." A few took an interest in their work and became proficient, but the greater number took no interest whatsoever. There was a general agreement among the large group of lumbermen who met at Hattiesburg to solve the labor shortage that Negro workers were undependable, shiftless, and that high wages merely made a bad situation worse.

In 1906 conditions were about the same as in 1905, perhaps worse. Even the offer of $1.50 to $2.00 per day made little change in the number of workers seeking unskilled jobs. The *American Lumberman* reported that mills were either closing down or shortening the work week because of the labor shortage. Frank Park, owner of the Mason Lumber Company, stated that Negroes and whites were always changing jobs and seemed uninterested in better working conditions. He had shortened the workday from twelve to ten hours with no reduction in pay and given the workers homes rent-free, yet no week passed without the disappearance of several Negroes.

Never had there been a greater demand for workers by lumbermen than during the first half of 1907. It was said that not a single mill on the Gulf Coast nor anywhere else in Mississippi had its full complement of workers. For short periods a mill might be fully manned; but following paydays most of the workers went on vacation and remained away as long as they had provisions to keep body and soul together.

Not everyone agreed with such a criticism of the colored worker. In 1895 one observer reported that Negroes were preferred over whites because of their superior strength and endurance. Actually, the Negro worker met and reacted to widely different types of treatment by his employers. The average Negro, unspoiled by education and life in the city, was, if dealt with properly, the best type of mill labor, as patient as an ox and as reliable as a steam engine. All he wanted was fair treatment, plenty of food, and a chance to frolic occasionally. Millmen in the yellow pine country who understood his needs had no trouble with the Negro.

An editorial writer in the *Southern Lumberman* in 1898 contended that in work which required a continuous muscular exertion no one else on earth could compete with the southern Negro. His muscular strength had given a name to a log-turning device in the sawmills, the "steam nigger."

S. S. Henry, a native southern lumberman, stated that as an employer of labor he had concluded that the Negro was the best sawmill worker in the world. He would perform as much work as two white men, and all he wanted was three square meals a day and his wages paid weekly. Supporting the testimony of Henry as to the quality of Negro labor were the observations of two German foresters, Volsprecht Riebel and Dr. Jentsch, who toured the southern pine country in 1906. They remarked that only the Negro was acclimated to the subtropical conditions.

His robust body stood the great heat as well as the swamp fevers.

One logging superintendent asserted in 1911 that for laying steel in tramroad construction, Negroes were better than either native white men or Mexicans. Negro crews would stay on the job if they were given plenty to eat, a place to shoot craps, a place to preach, and good living quarters.

Walter Barber, who was employed by the L. N. Dantzler Lumber Company as early as 1912 and became a mill superintendent and business manager, stated that many Negroes, although unable to read and write, performed complicated operations that required quick mental calculations and exceptional skills. In his experience the colored worker was reliable and dependable, though each individual required special handling. At times, rough treatment had to be used; at other times, kindness and appreciation.

The perennial shortage of labor and the desire to obtain workers of better quality, or workers of any description, caused lumbermen to look to the Old World as a possible source of millhands. Camp and Hinton reported that they had employed considerable numbers of Italians, but half of them were unsatisfactory. It was believed that Irish, Swedes, Norwegians, and Germans made reliable, consistent workers, and would advance themselves from laborers to property holders and good citizens. W. E. Guild of the Finkbine Lumber Company asserted that they were what the southern mills needed.

Actually no great number of European immigrants were ever employed in the Mississippi mills. Camp and Hinton appear to have had the greatest success. In 1907 the sawmill crew of the company was made up mostly of Italians.

The superintendent reported them to be reliable and readily satisfied with the wage scale. But Italian workers were said to be difficult to obtain. Their government refused to issue passports to those desiring to come to Mississippi because of previous mistreatment of Italian immigrants by Delta planters.

Only a few north Europeans, it seems, came to Mississippi. A few Danish workers obtained by the Finkbine Lumber Company remained only a short while. At D'Lo about sixty Norwegians were brought in. Perhaps the hot, humid climate caused the people of the northern climate to avoid the South. An immigration official stated that the Dutch and Scandinavians were beer-drinking people and would not migrate to states which maintained prohibition. This observation was borne out by a member of the Southern Pine Association, who reported that thirty Germans he brought from New Orleans came to him before the end of their first day at work to say they must quit: the reason —no beer!

In 1907 the depression came and the closing of the mills converted the labor shortage into a surplus. The immediate result of the panic was a reduction in wages of 20 to 25 per cent in some of the south Mississippi mills. According to George Smith, all the mills would have closed down had it not been for the fact that bond payments had to be met, taxes paid, and a supply of labor maintained. The severity of the depression is illustrated by the fact that in May nine mills out of twenty-three were closed down, while ten of the others were operating on a six-days-a-month schedule.

The scarcity of jobs brought a labor war between black and white employees. In the Camp and Hinton logging works, because wages were reduced the white workers left their jobs. When Negroes were hired as replacements, they were forced to quit by whites with firearms. In the Tatum mill near Hattiesburg, notices were posted warning Negroes to stay out of the mill and the owner was warned to discharge his colored employees, but Negroes continued to work without being molested.

One result of the cessation of mill activity in Mississippi was the complaint that towns were congested by the presence of large numbers of unemployed Negroes. Many feared that the hungry, idle millhands would get out of hand and cause an outbreak of violence. Calling the Negroes a menace to society, Governor James K. Vardaman ordered peace officers to keep them under close surveillance and to enforce strictly the vagrancy laws.

The depression of 1907-1908 brought a lowered standard of living for most mill workers and their families. Laborers, if employed at all, worked on a part-time basis at reduced wages. Many lumbermen would have closed down their mills entirely but for the necessity of providing subsistence needs for their employees. By December, 1907, cash payments had been discontinued in many mills. Those who were employed received only limited credit in company-owned commissaries.

No uniform or general wage scale had existed in the sawmills of Mississippi. W. S. Hagerty, a Hattiesburg lumberman, stated in 1913 that the manufacturers in his territory paid $1.10 per day for the labor of mature men, not because they could not afford more, but because they could get men to work for that wage. In his opinion wages were much too low, and labor costs had increased less over the years than any other factor of production. Because of low wages, poor housing, and long hours, Hagerty thought that the prevailing belief that every man with ability would rise above his surroundings did not apply to the sawmill worker. In other words, survival of the fittest did not operate among sawmill laborers; the workday of eleven or twelve hours, extended without relief through twelve months in the year at low wages, did not foster self-improvement. Hagerty asserted that there were exceptional individuals among the sawmill workers who ought to be given the opportunity to improve their lot.[8]

The workday, like wages, varied from one locality to another. In the nineties most of the mills operated not less than twelve hours a day; even a fifteen-hour day was not uncommon. Time lost through work stoppage was regained through the extension of the workday. From 1900 to 1912 the work week averaged between sixty-four and sixty-six hours per week. In many mills laborers worked thirteen hours a day to obtain a half-holiday on Saturday. For most of the mills the eleven-hour day was general in 1912.

Gradually over the years public sentiment had come to favor a reduction of hours. In 1912 the Mississippi legislature enacted the ten-hour workday for those engaged in manufacturing establishments except in cases of emergency. Labor which cut and hauled timber was exempted from the ten-hour law. Hagerty asserted that the reduction of work time would prove beneficial to both labor and manufacturer. Since labor would do as much work in ten as in eleven or twelve hours, wages need not be reduced. He believed that common laborers already were working for as little as was possible for survival. Philip S. Gardiner stated that men did as much work in ten as in twelve hours, and in 1906 his company cut the workday to ten hours. He called upon the manufacturers to assume responsibility for the welfare of their workers, pointing out that a betterment of social conditions improved the efficiency of labor. According to Gardiner the South had been so long in developing its economic system that its human resources had been neglected.

Two of labor's most serious grievances against lumbermen were the infrequency of paydays and payment in company script which could be used only at the store or commissary owned by the millman. Workers in some concerns were paid on the fifteenth of each month for the preceding month's labor. For labor performed in May workmen were paid on the fifteenth of the following June. Since most workmen were unable to save enough from each payday to tide them over for an entire month, they drew company script called coupons to supply themselves with subsistence. In many instances the worker traded the script to his employer for cash at a high rate of discount. In the vicinity of Hattiesburg in 1913, mills issued script or store checks to workers between monthly paydays. Holders of coupons could convert them into cash at a 10 per cent discount.

Southern sawmill workers generally lived in villages, towns, or small cities which grew up near the location of mills. On the railroads that traversed the pine country such towns were often composed almost entirely of sawmill workers and their families. According to Robert Fullerton of the Chicago Lumber and Coal Company, managing a yellow pine sawmill was extremely complex in comparison with managing a white pine mill. In the white pine region of the North, timber was floated long distances to cities already established where sawmills converted the timber into lumber. In the yellow pine country there was no parallel to Muskegon, Michigan, where 750,000,000 board feet were sawn in a six-months summer period. In the southern pine country the sawmill had to be located as conveniently to the growing timber as possible. Therefore a new town had to be built in the wilderness to house the workmen. Mill owners had to provide schools, churches, water supplies, medical services, department stores, and the usual accommodations of a modern town.

There is little evidence to indicate that lumbermen ever greatly abused the enormous power that was theirs in the sawmill towns. In Mississippi cases of actual peonage were virtually unknown. Competition among lumbermen for labor during most of the period was frequently a deterrent to the subjection of the worker to intolerable living conditions. J. E. Rhodes, a secretary of the Southern Pine Association, observed that with the massing of workers into great timber camps for months or even years and their congregation into little cities, the problems of the employer became complex and perplexing; he had to shoulder enormous responsibilities. In Mississippi, according to Rhodes, nearly 65 per cent of the people were in some way dependent upon the lumber industry; throughout the South the numbers of men employed by the lumber companies were so great that anything affecting their welfare touched either directly or indirectly the whole population. In the lumber country the mill owner was, Rhodes said, as much an overlord as the feudal baron of old. Hundreds of men and families looked to him for subsistence of body and mind and even soul. But the lumberman of 1915 did not want servile workers, nor did he desire to put the iron collar of ownership on those who depended upon him for a livelihood.

The sawmill worker was underpaid, lived in a nondescript house, and was able to obtain for himself only the plainest of food and clothing. Few were ever able to rise in the social and economic scale. Many were without ambition, easily satisfied with living only in the present, and little concerned about the future of themselves or their children. However, had the average worker been dissatisfied with the life he knew, he had little opportunity for improvement.

But there was a brighter side to living in the sawmill towns and villages. Schools and churches were superior to those the piney woods people had known in their rural existence. Life in the villages brought close neighbors and social interaction, and made available to many a greater quantity of material goods than they had known before.

In the "early days"—around the turn of the century— transient laborers, drifters, and lawless men flocked to the lumber towns. Gambling, drinking, and crimes of violence were common. Although the lumbermen exercised the power of law enforcement and soon got rid of most of the undesirables, many Negroes celebrated paydays by getting drunk, gambling, and not infrequently shooting one another.

In 1905 there were reported to be in Mississippi 527 logging camps containing 8,185 workers. An unnamed writer stated in 1913 that two-thirds of all the mills possessed logging camps. About one-half of these had semi-permanent houses while in the other half the workers were quartered in camp cars. The Finkbine Lumber Company, Edward Hines, the Westons, Eastman and Gardiner, and other firms built logging villages out in the forest. (See Plate X.) After the timber was cut in their vicinity, the villages were moved to new locations.

Philip Gardiner stated that in 1893 the men in his logging camp were lawless and engaged freely in the liquor traffic. In that camp there was neither school nor church. His company in 1902 made an effort to attract a better class of labor by providing schools and recreational facilities. At Wisner, the logging camp of the Eastman-Gardiner Company, a village of about eight hundred people was incorporated. Besides schools, the village was provided with churches and a Y.M.C.A.

A writer in the *Lumber Trade Journal* characterized the population of the camps as shifting; it never became identified as a group or formed any permanent society. Children of families that moved periodically from one location to another missed the memories of a place called home. Few of the boys or girls above fourteen years of age attended school; and most of them grew up in ignorance to follow the occupations of their parents.

In the logging villages during the long, hot summer days, children shouted and rolled in the dirt while their mothers sat gossiping in doorways or congregated at the creek with tubs and washboards. About six o'clock the hungry children drifted toward home, and the tired men came in from the woods. From each house issued appetizing odors of coffee, fried meat, and biscuits.

For the average camp the commissary was the center of the community where all food and clothes were purchased. Dry salt meat, flour, snuff, tobacco, overalls, brogan shoes, and cotton gloves were commodities in great demand. Here also was the telephone, the time clerk's office, the doctor's headquarters, and the post office. On rainy days men congregated at the commissary to swap tall tales about how they hauled logs out of bottomless hollows and over perilous grades.

Physically the people of the camp were usually well cared for. They had enough to eat and wear, and were content, knowing nothing better than life in the little two- or three-room shacks. The men worked hard, while the women were fully occupied in tending homes and babies. Occasionally there was excitement. Some workman who had put in ten hours a day for months would suddenly go on a grand spree and lose both job and reputation.

The lumber industry changed the way of life for the people of an entire region. The movement from the small farms and villages to the populated centers which grew up wherever large sawmills were erected altered attitudes, habits, and customs which had been characteristic of pioneer society. The new life in an industrial society conferred the benefits of improved health standards, educational opportunities, and living conditions, and greater opportunities to develop individual skills in various trades. But the sawmill economy had its darker aspects. The free life far from the sound of the steam whistle was gone; the discipline of a machine culture had arrived. With the disappearance of the forests, deer and other wild game passed away. Even to those who remained on the small farms, the new era brought great changes. They now produced crops for sale in the markets and ceased to provide for themselves the economic self-sufficiency that had been their pride in the pre-industrial era.

CHAPTER XVII

THE END AND A NEW BEGINNING

THE GROWING CONCENTRATION of lumber production in a few large mills produced uneasiness in the minds of many Mississippians, conjuring up the specter of monopoly in lumber and agriculture. They especially feared that millmen would develop profitable agriculture on the cutover lands after the timber was gone, and convert millions of acres in south Mississippi into large farms owned by absentee landlords. Many common folk believed that such a development would either doom their children to lives as tenant farmers or force them to leave the state. Politicians seeking the vote of the small farmers, therefore, advocated a limit on the size of landholdings and tried to break up large blocks already established.

In accordance with a provision of the state constitution of 1890 which conferred authority to prohibit nonresident aliens and corporations from acquiring unlimited property rights in land, the legislature in 1892 passed an act intended to limit the real estate holdings of manufacturing and banking corporations to $1,000,000 in value, and of other corporations to $250,000. The act provided that a corporation acquiring property from debt over and above the value allowed by law must dispose of the excess within a five-year period. In 1898 Governor Anselm McLaurin deplored all ownership of land by corporations and nonresident aliens. He maintained that those who farmed the land should be encouraged to become its owners. In 1900 McLaurin again advocated legislation that would prohibit nonresident aliens from acquiring land and would further restrict ownership of land by corporations.[1]

Despite opposition from the state government, the consolidation of large tracts of land continued in the years between 1890 and 1906. So long as the price of timberlands was cheap, the legal restriction on landholding in terms of dollars was no great handicap to the lumbermen. In the period 1900-1906, however, a number of large lumber firms moved into Mississippi. Their competition for virgin timberlands, most of which were already in the hands of a few large investors, soon pushed land prices up. This had the effect of decreasing the acreage of timber that might be legally acquired and so of placing a ceiling, if the laws regarding landholding were strictly enforced, upon the expansion of lumber firms. Under these circumstances, the Great Southern Lumber Company, and perhaps others, sought abolition of the million-dollar limit to landholding. Isaac Enochs, representing the Great Southern, contended in an address to the Mississippi legislature in 1906 that the basic policy of the state toward corporations was wrong. Industrialization and progress were synonymous, he said; limitations of this kind on land ownership hurt the lumber industry since the operation of a large mill required a heavy investment in expensive logroads and costly machinery. No one, according to Enochs, could afford to construct such a mill without control of sufficient timber to provide for continuous operation over a long period. After the timber supply was depleted, he argued, capital brought in by lumber corporations would benefit the state by flowing into secondary industries.[2]

A bill to raise the ceiling to $10,000,000 passed the legislature only to be vetoed by Governor James E. Vardaman, who explained his action as follows:

> *I cannot command words with which to express my disapproval of a law which permits corporations to hold such large amounts of land as is provided for in this section.*
>
> *The policy of our laws, for many years, be it said to the credit of her [sic] lawmakers, has been against the concentration of wealth in the hands of a few, for the elimination of trusts, and against the fostering of monopoly. This section, if it becomes a law, places the pine forests of South Mississippi within the hands of a few people to be exploited, used and employed for the benefit of a few against the interest of the many. If the legislature had undertaken in definite terms to create a lumber trust, and place the longleaf pine industry in the hands of a limited number of men to be used as their cupidity and self-interest might dictate, I cannot conceive of a method or a means by which it could have been more effectually done than is proposed to be done by the enactment of this law. In the material development of our state and the pecuniary well-being of its people we are all deeply interested, but I would warn you, in this age of money getting and cruel greed for gold, not to permit the glamor of great wealth to so affect your mental and moral vision that you will commit yourself to a policy which will necessarily work harm to our people in the future by making the rich richer and the poor poorer. The best product of any land is its men and women, and that is the happiest and most prosperous population where the wealth of the state is most equally divided between the people.[3]*

In 1906 the act of 1892 was revised to enable a manufacturing corporation to hold land necessary for its operations to an amount not exceeding $2,000,000 in value exclusive of buildings, machinery, and other fixtures.[4] The Meridian *Star* warned that Mississippi would rue the day her forests were destroyed by syndicates which had bought up all the land. Population of a permanent character was what the state needed, the editor maintained, and sawmills would not bring it. When the mills and timber were gone, the mill workers would go too, and the farmer, the man who was needed, would not settle on the land.

The extent to which statutory limitation actually curtailed the engrossment of large acreages by single owners is difficult to measure. In 1910 lumbermen pointed out that under the original act of 1892 (wherein the limitation was $1,000,000) it had been possible to acquire legally 1,333,000 acres. In 1909, because of the great increase in stumpage values, only 40,000 acres might be purchased by a single owner.

Early in 1910 agents of timber, cotton, banking, and other business interests of south Mississippi held a meeting to perfect an organization to work for revision of the corporation laws. The Great Southern Lumber Company, reputed to hold around seven billion board feet of timber in Mississippi, offered to construct at Columbia a duplicate of its great Louisiana mill if the state would abolish limitations on timberland ownership. Moreover the officers of the Interstate Trust Company, a lumberman's bank in New Orleans, let it be known that if the law was repealed they would purchase $6,000,000 worth of state bonds which had found no buyers.[5]

Edward Hines, one of the largest of the timberland owners, savagely attacked what he termed the state's "blue laws" against corporations. Already under indictment for violation of the landholding law, Hines contended that the blue laws prevented lumbermen from competing with business in sister states. Hines asserted that he planned to build a chain of sawmills through his hundreds of thousands of acres of timberland. It was rumored that he and F. L. Peck, stockholders along with the Goodyears in the Mobile, Jackson, and Kansas City Railroad, might, if they could legally acquire more timberland, build a railroad north of the Ohio River to reduce freight rates.

In 1909 the Hines interest had been sued by the state for violation of the statute prohibiting acquisition of land exceeding $2,000,000 in value. The value of his 241,000 or more acres of timber was estimated to be $8,000,000. Supported by the business community, Hines tried to obtain a compromise. The case attracted a great deal of attention, particularly in the counties where Hines was a large landowner. In Pearl River County, for example, six thousand citizens signed a petition asking that the case against him be dropped. As a condition of the compromise proposed by the citizens, Hines was to agree to cut and remove his timber within a reasonable time, then sell the cutover lands to single purchasers in blocks of forty to five hundred acres.

Hines's attorneys asserted that all of his lands except a small tract valued at $25,000 had been acquired prior to passage of the 1906 act. Even if this had not been true, the law did not apply to the Edward Hines Lumber Company, they argued, because it was not a Mississippi corporation; Mississippi could not penalize by forfeiture of charter and confiscation of lands a company incorporated under the laws of another state. Eventually the court ruled that foreign corporations were not subject to the provisions of the act. After this decision the limitation on ownership of timberlands had no effect in Mississippi.[6]

Almost all lumbermen contended that the system of state and county taxation employed in Mississippi was unfair to owners of forest lands. Timberlands and farmlands were taxed similarly; the amount of the tax was determined by such varying factors as the assessed valuation of the land and the tax levy. Lumbermen argued that an annual ad valorem tax was unrealistic, for forest lands, unlike agricultural lands, yielded no annual return. Timberland owners favored instead a small land tax together with a severance tax on timber. They argued that a high annual tax on land containing timber would compel the owner to fell the trees as quickly as possible; otherwise the accumulation of taxes over a number of years could exceed the value of the timber.

After 1900 stumpage prices continued to increase and so did the assessments placed on timberland by the state and local governments. In Perry County in south Mississippi, for instance, the estimate of board feet per acre, valuation of timber, and tax levy brought an increasingly higher tax on each acre of forest land. In Perry County an acre of virgin timber formerly valued at $8 to $10 was suddenly raised to $15 to $18. Forest lands in Harrison County also rose from $1.25 per acre to $5.00 in 1905 and $12 to $15 in 1909.[7]

The tax burden in all south Mississippi counties rested mainly on land and timber. In 1908 the maximum levy authorized by state statutes was eighteen mills. But soon the demand for better schools and roads brought an increase in assessed valuations, tax levies, and estimated quantity of board feet per acre. The cost of services provided by local governments was rising at a time when the taxable wealth represented by virgin timber was fast disappearing. The result in most cases was a higher and higher tax rate on the forested lands remaining.[8]

The upward trend in taxation is exemplified by the action of the board of supervisors of Pearl River County in relation to the Blodgett holdings. During a twelve-month period the valuation of the Blodgett property was raised from $808,000 to $1,334,665, increasing the annual tax by $15,880.50. This action caused John Blodgett to institute court action against Pearl River County for what he termed an unjust act. He asserted that such high assessments were harmful to the state because they would lead to quick cutting; the resulting large areas of cutover land would be a drug on the market. Edward Hines contended in 1908 that the tax rate was unreasonable. He paid around $50,000 annually on 241,000 acres in Mississippi, or approximately twenty cents on each acre, whereas on his Canadian holdings of 500,000,000 board feet of stumpage taxes were only $300. Because the Hines company left a few small trees standing on its lands after cutting operations, the board of supervisors of Pearl River County refused to reclassify them as cutover lands for tax purposes. Hines then had all his lands entirely cleared of pine trees in order to obtain reduced taxes. Thus a short-sighted policy of county officials contributed to complete destruction of the forests.[9]

If statements of Mississippi lumbermen were to be taken at face value, their quick cutting of timber and general lack of interest in reforestation were directly caused by high taxes. Actually they had other reasons of even greater importance for the quick cutting of timber. In the first place, a large percentage of the timberland they owned stood as security for their debts, and continuous operation of their mills was necessary to meet payments of interest and principal. In the second place, although millmen almost never mentioned this fact, the volume of standing timber on an acre was bound to decrease with each passing year. The rate of growth of virgin forests was extremely slow, and each tree destroyed by fire or other hazards brought a decrease in the volume of board feet per acre.

Lumbering operations denuded hundreds of thousands of acres annually. Naturally, the subsequent use of these lands was of paramount importance to the people of the state. The owners and general public believed that reforestation was at best a long-drawn-out process and likely to be expensive. Moreover, the tax system designed for agricultural tenure did not encourage lumbermen to reforest their barren acres.

In 1908 a report compiled by United States foresters revealed that more than half of the longleaf pineland of Mississippi had already been converted into stumps. At this rate of cutting, the longleaf lumber industry would be reduced in twenty-five years to the few large mill companies which owned most of the timberland; within that time the supply of virgin timber would be exhausted. In the opinion of the foresters, less than half the longleaf pinelands would ever be suitable for farming; in the hilly sections west of the Pearl River and in the coast region, the arable area would be far less than half.[10]

The conservation movements that spread throughout the United States in the first decade of the twentieth century gained no secure foothold in Mississippi. Lumbermen apparently believed that timber having the quality of virgin forests could never be reproduced. Enochs considered cutover lands as worth more for growing cotton than for timber. George S. Gardiner, president of the Eastman-Gardiner Lumber Company, stated that lumbermen could not afford to reforest their lands because of high taxation. Taxation aside, he believed the longleaf pine country had greater value for agriculture than for growing trees. C. S. Butterfield asserted that reforestation was impractical as long as the unjust system of land taxation prevailed.

Senator J. D. Donald of the Mississippi state legislature called upon lumbermen to support the enactment of a law to prohibit cutting trees below fourteen inches in diameter. As part of his plan, lands in growing forest would be classed as cutover lands, and timber would be taxed only when harvested. Mississippi lumbermen thought highly of Donald's tax proposal but not of the restriction on timber cutting. They argued that limiting lumber operations would be disastrous to current business relations and would put the small millmen out of business.

C. W. Goodyear of the Great Southern Lumber Company, for example, strongly opposed all restrictions on the size of trees to be cut on the grounds that the state would establish the right to interfere with private property if it regulated the cutting of timber. He maintained that lumbermen had gone into business believing that there was stability in law governing land tenure in Mississippi. So long as he did not interfere with the rights of the rest of the community, he was entitled to buy a piece of land and use it as he wished. Timber was, according to Goodyear, no more than a crop, and the person who bought it had the right to make a profit from it if he could.[11]

Robert Fullerton, on the other hand, contended that it was a crime to cut a tree less than ten inches in diameter, since yellow pine could not be produced artificially. Any state restriction on cutting timber invaded vested rights, however, and legislation to that end was therefore undesirable. He apparently favored voluntary control of cutting by the individual lumberman.

The ownership of thousands of acres of cutover lands constituted a major problem for lumbermen. Many viewed such lands as worthless and regarded their sale at any price as an unexpected windfall. Others were convinced that the pinelands were potentially useful for agriculture. Whatever their views on this point, most lumbermen were happy to sell denuded lands to prospective farmers. To hold them meant an annual financial drain for taxes. Lumbermen were either unwilling or unable to introduce reforestation.

In the years 1909-1914 a general movement was initiated by lumbermen to dispose of their millions of acres of cutover lands. Land agents of the lumber companies sold thousands of acres to prospective farmers in the North. Lumbermen themselves cleared large tracts and established demonstration farms.

The clearing of land and establishment of farms by lumbermen apparently convinced many people in Mississippi that the pine country would one day become a prosperous agricultural region and led them to fear that lumbermen might sell their holdings in large blocks to agricultural corporations which would transform the pinelands into large farms. The result of this fear was an act by the legislature in 1912 which provided that a corporation could in no case buy or acquire farmland for agricultural purposes. It might, however, be lessor or lessee of as much as ten thousand acres of such land for a period not longer than twenty years. It might also obtain farmlands through foreclosure in collection of debts, but property so acquired must be disposed of within a twenty-year period. Six years later, after the wave of immigration and the movements of lumber companies to develop their lands had largely played out, the 1912 act was amended to encourage corporate owners of cutover lands to develop their holdings into farms. Subject to statutory limitations, corporations were authorized to develop, improve, and cultivate wild cut-over pinelands and sell them to any buyer other than a corporation. A corporation could cultivate only 20 per cent of its lands in a single county, or, if these measured less than 1,000 acres, not more than 50 per cent.[12]

Farming ventures of lumbermen in almost every instance proved to be unprofitable. For most of the land-buying colonists, clearing the land of stumps and attempting to grow crops on the thin pine soils brought only hard work, disillusionment, and tragic failure. Stump removal required backbreaking toil, and the long hot summer season sapped energy and vitality. Even after a few acres had been cleared, the land failed to produce the bountiful crops promised by newspaper articles, pamphlets, and the slick tongue of land salesmen. After a year or so of unprofitable toil, settlers usually packed their small belongings and re-turned, reduced in worldly goods, to the land whence they had come. The small clearings left behind grew up in broom sedge to bear mute testimony to the blasted hopes of those who had tried to build farms in a country designed by nature for growing pine trees.[13]

By 1920 the once common belief that most of the cut-over land would one day be converted into prosperous farms was fast disappearing. The millions of denuded acres, growing in number each year, posed an increasingly seri-ous problem not only to the owners but also to the state. Yet the old ways of wasteful logging continued; the land was swept clean of young trees by woods fires in every spring season; and only a few lumbermen, almost all of them natives, took steps to reforest their cutover lands. Posey Howell, a minor official of the L. N. Dantzler Lumber Company, began the practice of leaving seed trees upon company lands. Howell eventually convinced his employers that reforestation was both practicable and potentially profitable. The Westons also were pioneers in growing trees on their thousands of idle acres. The Tatums, who came to Mississippi from Tennessee, began early to cut only mature timber. W. S. F. Tatum, founder of the company, discovered through experiments that harvesting immature timber did not increase profits. Randoph Batson, another native lumberman, made some effort to reforest his eighty thousand acres. The Great Southern Lumber Company turned to reforestation after its expensive agricultural experiments failed to produce bountiful crops and high grade livestock. In the twenties the company began planting what is now claimed to be the largest man-made forest in the world.[14] But the great majority of lumbermen, particularly those who came from outside the South, stripped the land of its forests and later moved to the Pacific Coast.

In the twenties there was a gradual awakening to the fact that millions of acres in south Mississippi would perhaps always be best suited to growing pine trees. This change of attitude was reflected in an act of the legislature in 1926 providing for the establishment of a forestry commission, making a small appropriation for forest fire protection, and granting the Federal Government the right to acquire a maximum of twenty-five thousand acres to establish a national forest.[15] With each passing year additional acreages of cutover land were brought under fire protection. In the thirties the Federal Government acquired 1,245,000 acres of cutover lands, paying about three dollars per acre for much of the same acreage which as virgin timberland had been sold by the United States to private individuals for $1.25 an acre in the 1880's. The Federal Government was at long last doing in Mississippi what had been recommended by Carl Schurz sixty years earlier. In the late thirties the work of both state and national governments in reforesting the land began to bear fruit. In many localities young green pine forests began to appear upon the bare ridges and open flat lands. Slowly, and in many cases unwillingly, the people changed their age-old custom of burning off the wire grass in the early spring.

That a tax system designed for agricultural tenure was unsuited to the needs of forest lands was eventually recognized in Mississippi. The legislature in 1940 gave the holders of timberland all that they asked for, perhaps too much. To encourage reforestation, the ad valorem tax on standing timber was repealed. In its place was substituted a small severance tax to be collected when the timber was cut. The law fixed a low assessment on the land to be paid annually.[16]

Capital accumulated in the period of the big mills and virgin timber was used to develop secondary forest industries. In 1911 one of the early pulp mills in the South was constructed to utilize the waste materials of the Dantzler Lumber Company's mills at Moss Point; later it became a heavy consumer of small second growth pine. At Laurel the Eastman-Gardiner Lumber Company, after exhausting their supply of virgin pine, turned to the manufacture of hardwood lumber. Also at Laurel the Masonite Corporation established one of the first factories in the world for converting pine and other woods into fiberboard, shingles, and other building materials. Creosote factories or wood preservation plants were established in a number of localities. It may be noted incidentally that many if not all south Mississippi banks were started by capital derived either directly or indirectly from forest industries; that Eastman-Gardiner at Laurel and the Lamptons at Magnolia came to operate cotton mills; and that the Finkbine Lumber Company established at Wiggins one of the largest pickle canning plants in the world.

Production of lumber in Mississippi in the years 1915-1932 ranged from a high of 3,259,194,000 board feet in 1925 to a low of 800,000,000 in 1932. While in the next twenty years annual production averaged less than it had in the period 1915-1932, Mississippi in 1956 had 1,087 saw and planing mills, forty-seven furniture factories, and twenty mills producing pulp.[17]

The importance of forest industries to the state's economy is borne out by the report of the state tax commissioner for 1959. Timber products harvested in Mississippi came to 1,909,555,000 board feet. Of this amount the twenty-five counties comprising the old longleaf pine belt produced 866,573,000 board feet, almost one-half of the total. For the same period 208,698 tons of distillate and 17,197 barrels of gum were harvested.

The future of forest industries in Mississippi is bright. Forests cover more area than the combined acreage of all other land uses in the state. According to the third forest survey made in 1958, forests now occupy 57 per cent or 17,200,000 acres. Annually over the past ten year period, 70,000 acres of submarginal agricultural land have reverted to forest land.

APPENDIX

1. Manuscript Collections

As yet comparatively few records relating to the economic history of south Mississippi have been collected. For this reason no single body of primary sources satisfactorily covers the forest products industries of that region for the period between 1840 and 1915. Furthermore, there are intervals within this time span in which historical information on this subject is almost entirely lacking. Nevertheless, several collections of documents indispensable to the history of lumbering and naval stores are available.

Perhaps the most valuable of these collections is the Calvin Taylor and Family Papers, which are in the Louisiana State University library. Virtually every phase of lumbering from the gathering of raw materials to the sale of the finished product is commented upon in the journals, letters and diaries of Calvin Taylor and his brother, Sereno.

Descriptions of sawmills and logging operations on the Mississippi Gulf Coast during the 1850's are included in the B. L. C. Wailes Diary, Notes in the Field No. 4, deposited in the Mississippi Department of Archives and History, Jackson.

Two collections in the Lumber Archives of the University of Mississippi are of unusual value for the early twentieth century. The Weston Collection consists of business and personal correspondence, business records, and memoirs for the period 1911-1945.

A few letters written by Henry Weston, founder of the company, relate to the ante-bellum era. There is also his brief but important manuscript history of the company from 1845 to 1900.

The other important collection at the University of Mississippi is the L. N. Dantzler Collection, a very large body of correspondence and business records of many types. These papers are the best source of information about the Gulf Coast lumber industry during the first quarter of the 1900's. In addition, a manuscript history of the company written by Walter Barber is based upon company records, and, in general, is reliable.

Charles Shotts's unpublished "History of the Mississippi Forestry Commission" (1954), in the possession of the author, is a useful account of the activities of pioneers in the reforestation movement.

2. U. S. Government Documents

Statistics compiled by the Bureau of the Census provide a general picture of forest industries during the nineteenth century. Schedule No. 5 of the Seventh Census 1850 and the Eighth Census 1860 relate to the ante-bellum era. Similar information for the postwar decades is provided by Schedule No. 4, Ninth Census 1870 and Special Schedule of Manufactures No. 3 of the Tenth Census 1880. Data concerning slave labor in the industry can be found in the Slave Inhabitants, Schedule No. 2, of the Seventh Census 1850 and of the Eighth Census 1860.

Publications of the Federal Government are the best sources of information on law, court actions, and general governmental activities. U. S. Congress, House Committee on Ways and Means, *Tariff Hearings* (1908-1909), Vol. 3 contains information on lumber distribution and land taxation which is not available in any other source. U. S. Bureau of Corporations, *The Lumber Industry*, Vol. 4 is a good analysis of the monopolistic aspects of the lumber industry. U. S. Congress *Report on the Internal Commerce* (49th Cong., 2nd Sess., House Exec. Doc. 7, Serial 2476), Charles Mohr's *Timber Pines of the United States* (U. S. Bureau of Forestry Bull. No. 13, 1896), Charles S. Sargent's *Report on the Forest Trees of North America (Exclusive of Mexico)* and A. W. Schorger and H. S. Betts's *The Naval Stores Industry* (U. S. Forest Service Bull. 229, 1945) contain data of the forest industries of Mississippi which are not available from other sources. The reports compiled by Sargent and Mohr are the most reliable on the status of the forest industries during 1880 and the period from 1885 to 1892.

Report of the Commissioner of the General Land Office (1852 to

1900) is one of the best sources of information on the disposal of government timberlands as well as related subjects. Other primary sources dealing with the land policies of the Federal Government are *U. S. Statutes at Large,* IX, X, XIV, XIX, XXI, XXV, XXVI. Thomas Donaldson's *The Public Domain: Its History, with Statistics,* 97th Cong., 2nd Sess., House Misc. Doc. No. 45, Pt. 4, has valuable facts pertaining to the land policies of the Federal Government for the period before 1884. The *Congressional Record* for the years 1876 to 1910 is an excellent source for information on the attitudes of Mississippi representatives in Congress toward Federal land policies. *Interstate Commerce Commission Reports* for the years 1904, 1905, 1909, and 1912 are the most authoritative sources on the problems of lumber as well as log transportation. The *Reports* also contain data on the development of the yellow pine industry in the South.

3. Publications of the State of Mississippi

Mississippi Laws, covering the period 1852 to 1930, contains legislation dealing with land disposal, land tenure, land taxation, and forestry. Various acts concerning labor are found in the session laws of 1902, 1904, 1908, 1912, and 1914. The *Mississippi Senate* and *Mississippi House Journals* provide an outline of the legislative history of bills and resolutions. *Mississippi House Journal* (1886) contains a lengthy investigation of fraudulent land sales made by the state. In the *Mississippi House* and *Mississippi Senate Journals* from 1899 to 1900 are messages of governors which express their views on land tenure. The reports of the Swampland Commissioner and the Commissioner of State Lands, from 1877 to 1902, have information not available elsewhere on the management and the disposal of lands. *Mississippi Reports* and *Southern Reporter* interpret laws dealing with such matters as land and labor legislation. The publications of various state departments and agencies contain much data related to lumbering, forestry, and transportation. The most useful sources are the *Biennial Reports of the Railroad Commission* for the period 1910 to 1916, and the *Annual Reports of the Forestry Commission.* The best single work on forest industries and land use is in E. N. Lowe's *Forest Conditions in Mississippi* (Jackson, 1908).

4. Periodicals

The most important sources of information about all phases of the lumber industry are the trade journals. Two such publications, the *Lumber Trade Journal* and the *Southern Lumberman,* devote much attention to the southern yellow pine industry. The *American Lumberman* and the *St. Louis Lumberman* are also valuable. For the period between 1880 and 1900 the *Northwestern Lumberman* and the *Southern Lumberman* provide the most useful accounts of Mississippi lumbering.

Other journals of lesser value are *Lumberman's Review, Lumber World Review, Lumber Review,* and the *Timberman.* John F. H. Claiborne's two articles, "Rough Riding Down South," *Harper's New Monthly Magazine,* XXX (June 6, 1862), and "A Trip Through the Piney Woods," *Publications of the Mississippi Historical Society,* IX (1906), are excellent descriptions of the longleaf pine country and the people who lived in the region.

Three short articles in *De Bow's Review,* I (1846), XXIV (1858) and XXVII (1859) describe lumbering on the Mississippi Gulf Coast. P. L. Buttrick's "Commercial Uses of the Longleaf Pine," *American Forestry,* XXV (Sept., 1915), is a brief summary of the uses of the longleaf pine and various phases of forest industries. Another unsigned article in the same issue of *American Forestry* also discusses the longleaf pine industry.

Paul Wallace Gates's "Federal Land Policy in the South: 1866-1888," *Journal of Southern History,* VI (Aug., 1940), is the standard study on Federal land policy in the South. Carl V. Ostrom's "History of the Gum Naval Stores Industry," *The Chemurgic Digest,* IV, No. 12 (July 15, 1945), deals exclusively with naval stores.

5. *Histories, Biographies, and Monographs*

Very few books have been published on the history of the forest industries in the longleaf pine belt of Mississippi. One of the few secondary works dealing primarily with lumbering is *American Lumbermen: the Personal History and Public Achievements of One Hundred Eminent Lumbermen of the United States,* 3 vols. (Chicago, 1905-1906). This publication is a series of biographical sketches of lumbermen, who probably supplied the information themselves, but only a few of the essays are about Mississippi lumbermen. A brief but excellent account of the lumber industry and the state policies in respect to corporations and land tenure is a pamphlet written by the lumberman Isaac C. Enochs, entitled *Corporate Holding of Timberlands* (Syracuse, 1906). Another pamphlet by Willard F. Bond, *Life of Jim Broom* (Cleveland, 1950), discusses the economy of the pine country and rafting of logs on Red and Black creeks and the Pascagoula River. Cyril Edward Cain's *Four Centuries on the Pascagoula* (Starkville, 1953), based upon his personal experiences, is a good description of log rafting and the economy of the inhabitants of southeast Mississippi.

Two scholarly works, Robert F. Fries's *Empire in Pine, the Story of Lumbering in Wisconsin: 1830-1900* (Madison, 1951) and Agnes M. Larson's *History of the White Pine Industry in Minnesota* (Minneapolis, 1949), give pertinent facts on Federal land policies and technological developments in lumber manufacturing. James Defebaugh's *History of the American Lumber Industry*, 2 vols. (Chicago, 1906) has little material concerning the lumber industry in Mississippi and other southern states. It does, however, throw light on the tariff policies which affected lumbering throughout the United States. Henry Disston and Sons' *The Saw in History* (Philadelphia, 1915) purports to be a history of the saw, but it is highly inaccurate and unreliable. The only book dealing with naval stores is Thomas Gamble's *Naval Stores: History, Production, Distribution, and Consumption* (Savannah, 1921). Gamble was editor of the *Naval Stores Review* and had a firsthand acquaintance with the various phases of the industry. His book has considerable merit as a broad treatment of the naval stores industry, but it is deficient with respect to the industry in Mississippi.

Peter J. Hamilton's *Colonial Mobile* (Boston, 1910) and his *Mobile Under Five Flags* (Mobile, 1913) contain a few interesting facts on forest industries along the Gulf Coast during the colonial period. Also Lewis C. Gray's *History of Agriculture in the Southern United States to 1860*, 2 vols. (New York, 1941) briefly mentions the existence of forest industries on the Gulf Coast. For a good discussion of agriculture and soils and a brief, reliable account of early forest industries, there are B. L. C. Wailes's *Report on the Agriculture and Geology of Mississippi* (Jackson, 1854) and Eugene Hilgard's *Report on the Geology and Agriculture of Mississippi* (Jackson, 1860).

Useful accounts on the economy of the inhabitants of the longleaf pine country are John H. Lang's *History of Harrison County* (Gulfport, 1936), Luke Ward Conerly's *History of Pike County* (Nashville, 1909), L. A. Besancon's *Besancon's Annual Register of the State of Mississippi* (Natchez, 1838), William H. Sparks's *Memories of Fifty Years* (Macon, Ga., 1872), and Etienne Maxson's *Progress of the Races* (Washington, 1930). The most sympathetic treatment of the type of people who inhabited the pine country is Frank L. Owsley's *Plain Folk of the Old South* (Baton Rouge, 1949).

The relation of the lumber industry to railroad development and transportation is traced in the excellent monograph by James Hutton

Lemly entitled *The Gulf, Mobile, and Ohio* (Homewood, Ill., 1953). F. W. Taussig's *The Tariff History of the United States* (New York, 1923) has hardly anything to say about the effect of tariffs on the yellow pine industry.

6. *Newspapers*

Newspaper sources for the early period of Gulf Coast lumbering are scanty. The Pascagoula (Miss.) *Democrat Star* is the best source for the period 1874-1910. The New Orleans (La.) *Daily Crescent* is valuable for the years 1855 to 1860. Around the turn of the century, the Biloxi (Miss.) *Daily Herald* published significant articles on forest industries.

7. *Interviews with Lumbermen*

The recollections of persons long associated with forest industries are important sources of information obtainable nowhere else. The depositions in possession of the author of John Gary, Joseph Simpson, Dennis Smith and others employed for forty years or more in naval stores production, for example, are the best sources on working and living conditions of the turpentine workers. Walter Barber, Wes Tatum, and Mayers Dantzler contributed much valuable information on lumbering, and William Batson, surveyor and land owner, provided significant data on the results of the state and Federal land policies in the longleaf pine country. John F. Hickman and Daniel G. McQuagge, old-time rafters and loggers, also gave firsthand descriptions of all aspects of lumbering along the Gulf Coast.

NOTES

CHAPTER I

[1]The following references are frequently drawn upon: Charles Mohr, *Timber Pines of the Southern United States*, U. S. Bureau of Forestry, Bull. No. 13 (Washington 1896) 30, 42-43, map opposite 30; E. N. Lowe, *Forest Conditions in Mississippi*, U. S. Forest Service in co-operation with Mississippi Geological Survey, Bull. No. 11 (Jackson 1913) 38-39, 83, 97-99, 161; Eugene Hilgard, *Report on the Geology and Agriculture of Mississippi* (Jackson 1860) 340-349, 370-373, 368, 366, 364, 356; Charles S. Sargent, *Report on the Forest Trees of North America (Exclusive of Mexico)* 47th Cong., 2nd Sess., House Misc. Doc. 42, Vol. 9 (1883) 532; observations of the author in Stone County, Mississippi, 1923.

[2]John F. H. Claiborne, "A Trip Through the Piney Woods," *Publications of the Mississippi Historical Society* IX (1906) 523.

[3]*De Bow's Review* I (1846) 254.

[4]Soils surveys made by the U. S. Department of Agriculture in co-operation with the Mississippi Geological Survey in the counties of Jackson, Harrison, Hancock, and George.

[5]Statement made to the author by Joe Simpson, June 3, 1954. Simpson, born and reared in the longleaf pine country of Georgia, was for thirty years manager of the turpentine operations of the Finkbine Lumber Company in Simpson, Rankin, Stone, and Pearl River counties, Mississippi. Statement made to the author by Mrs. W. O. Finkbine, Sept. 13, 1953. Mrs. Finkbine, wife of one of the owners of the Finkbine Lumber Company, stated that the pine trees in Rankin and Smith counties were exceptionally large, but were overripe, and consequently made a poor grade of lumber. Statement made to the author by John Cross, Jan. 4, 1953. Cross, a supervisor for the U. S. Forest Service, has been connected with the naval stores industry for more than twenty years in Mississippi, Alabama, and east Louisiana.

[6]Rand McNally, *Railroad Commissioners' Map of Mississippi*, in Mississippi State Land Office, Jackson, Miss.; Etienne Maxson, *Progress of the Races* (Washington 1930) 10. Maxson's father, a Negro slave, operated a cotton gin near the mouth of the Pearl River at a landing known as Cotton Gin Port. Cotton was brought down the river in boats from the back country to be processed at the gin and shipped thence to New Orleans. Besides the damage it inflicted on the Moss Point mills, the storm of September, 1906, destroyed in some places around 70 per cent of the timber. Pascagoula (Miss.) *Democrat Star*, Oct. 1, 1906. Cain, who describes briefly some twenty severe hurricanes on the Mississippi coast in the years 1699-1916, says that the September storm in 1906 was the most destructive since 1893. See Cyril Edward Cain, *Four Centuries on the Pascagoula* (Starkville, Miss. 1953) 196-204.

[7]United States Census, Tenth Census 1880, [I] *Population,* 67-68; L. A. Besancon, *Besancon's Annual Register of the State of Mississippi* (Natchez 1838) 190.

[8]Notes in the Field No. 4, Diary of B. L. C. Wailes, 1852, 14, in State Department of Archives and History, Jackson, Miss. This work is hereinafter cited as Wailes, Notes.

[9]Besancon, *op. cit.* 190.

[10]Tenth Census 1880, [I] *Population,* 67-68.

[11]Benjamin L. C. Wailes, *Report on the Agriculture and Geology of Mississippi* (Jackson 1854) 200-205. This work is hereinafter cited as Wailes, *Agriculture.* Hilgard, *op. cit.* 366-367, 384.

[12]Raymond (Miss.) *South-Western Farmer*, Dec. 6, 1844, 201; Frank L. Owsley, *Plain Folk of the Old South* (Baton Rouge 1949) 72, 37-38.

[13]William H. Sparks, *The Memories of Fifty Years* (Macon, Ga. 1872) 331-332.

[14]Maxson, *op. cit.* 9; Raymond *South-Western Farmer*, May 18, 1843, 137.

[15]Statement made to the author July 28, 1953, by Wilmer Griffin, who was born in 1866 and lived in the home of his grandfather, William Griffin, while attending school in Moss Point, Mississippi. According to Griffin, his grandfather came to Perry County, Mississippi, from Georgia with his parents in 1808. Willard F. Bond, *The Life of Jim Broom* (Cleveland, Miss. 1950) 4; Wailes, Notes, 77.

[16]Statements of Griffin and Clyde Ward; the latter made Nov. 22, 1953.

[17]Owsley, *op. cit.* 67-72.

[18]Statement to the author, June 26, 1953, by Calvin E. Dees, the grandson of A. W. Ramsey. Dees's knowledge concerning his grandfather came from his grandmother, Ramsey's wife.

[19]John F. H. Claiborne, "Rough Riding Down South," *Harper's New Monthly Magazine* XXX (June 6, 1862) 29.

[20]Pascagoula *Democrat Star*, April 2, 1892. As a boy in the late ante-bellum period Mark A. Dees worked in a sawmill owned by his father. In his later years he wrote many articles for the local newspapers on the lumber industry at the mouth of the Pascagoula River. He remembered that with the disappearance of cane and reed the piney woods people were forced to become timber workers.

[21]Hilgard, *op. cit.* 361.

Chapter II

[1]Peter J. Hamilton, *Mobile Under Five Flags* (Mobile 1913) 68, 153, 165, 178; Dunbar Rowland, *Mississippi Provincial Archives, 1704-1743* (Jackson, Miss. 1932) III, 573; Peter J. Hamilton, *Colonial Mobile* (Boston 1910) 241, 302, 326, 68; C. N. Howard, "Some Economic Aspects of British West Florida, 1763-1768," *Journal of Southern History* VI (May 1940) 217; Lewis Cecil Gray, *History of Agriculture in the Southern United States to 1860* (New York 1941) I, 74, 75, 159, 334.

[2]Robert F. Fries, *Empire in Pine, the Story of Lumbering in Wisconsin, 1830-1900* (Madison 1951) 60-61; Henry Disston and Sons, *The Saw in History* (Philadelphia 1915) 3; *Southern Lumberman*, Aug. 1882, 2.

[3]James Defebaugh, *History of the American Lumber Industry* (Chicago 1906) I, 529.

[4]Pascagoula *Democrat Star*, Nov. 24, 1893; Sixth Census 1840, *Compendium* 236.

[5]Diary of a Lumberman, 1865-1868, unpublished manuscript deposited in the Library of Tulane University.

[6]*Timberman*, May 24, 1890, 48; *Lumber Trade Journal*, Aug. 15, 1901, 19; *Southern Lumberman*, Dec. 1, 1895, 11. The information concerning Henry Weston is contained in a typewritten statement of twelve pages in possession of J. Roland Weston of Bay St. Louis, Mississippi. Other Weston papers include cash books, journals of lumber accounts, land records, and miscellaneous papers. Most of the records are in the lumber archives of the University of Mississippi.

[7]Seventh Census 1850, Schedule No. 5, *Products of Industry*, manuscript returns for Clarke, Copiah, Covington, Greene, Hancock, Harrison, Jackson, Jasper, Jones, Lawrence, Marion, Perry, Pike, Simpson, Smith, and Wayne counties. Seventh Census 1850, Schedule No. 1, *Free Inhabitants*, manuscript returns for Clarke, Copiah, Covington, Greene, Hancock, Harrison, Jackson, Jasper, Jones, Lawrence, Marion, Perry, Pike, Simpson, Smith, and Wayne counties. Apparently the census marshall was careless with decimals and zeros in his figures concerning prices of lumber in Hancock County. For instance, the 1,500,000 board feet of lumber produced by the Wingate Mill was valued at $150,000, or $100 per thousand board feet, a preposterous price. In Harrison County, which bordered on Hancock, mills received on an average $6 to $8 per thousand board feet, and in Jackson County $5 to $11. Wailes, Notes, 34; *De Bow's Review* XXVII (1859) 596; Jackson *Mississippian*, July 1, 1857.

[8]Pascagoula *Democrat Star*, Nov. 24, 1893.

[9]L. N. Dantzler Lumber Company Records, Miscellaneous Papers. The heavy rectangular frame which had served to lend stability to the sash saw was eliminated. The muley saw was held in place at the lower end by a crosshead mounted in guides to which the end of the saw and the connecting rod were attached. On the upper end of the saw was another still lighter crosshead working in guides on a pendant support. Immediately above and below the log, the saw blade was held and guided by lateral supports of wood. The muley saw made from 200 to 400 strokes per minute. Disston, *op. cit.* 12-13; *Southern Lumberman*, Aug. 1, 1882, 1.

[10]L. N. Dantzler Lumber Company Records, Abstract Book No. 1040, 26; Pascagoula *Democrat Star*, Nov. 24, 1893; statement of Griffin; Fries, *op. cit.* 61; Walter Barber, History of the L. N. Dantzler Lumber Company, Perkinston, Mississippi, 2-3. This history traces briefly the story of the firm from its beginning in 1858 to 1953. Barber has been associated with the company since 1912.

[11]Bostick H. Breland, unpublished manuscript in a collection of essays on various subjects, in the possession of Houston Breland, Wiggins, Mississippi.

[12]Seventh Census 1850, Schedule No. 5; Pascagoula *Democrat Star*, Nov. 24, 1893.

[13]Disston, *op. cit.* 36; Fries, *op. cit.* 61; *Southern Lumberman*, Aug. 1, 1882, 2.

[14]Pascagoula *Democrat Star*, April 3, 1900; Nov. 24, 1893.

[15]Wailes, Notes, 72.

[16]Luke Ward Conerly, *History of Pike County* (Nashville 1909) 130; *American Lumbermen: The Personal History, Public and Business Achievements of One Hundred Eminent Lumbermen of the United States* (Chicago 1905, 1906) I, 367. This work is hereinafter cited as *American Lumbermen: History.*

Chapter III

[1]Canton (Miss.) *Independent Democrat*, July 1, 1848; Pascagoula *Democrat Star*, Dec. 11, 1891, July 5, 1889; Gordon Davis to Calvin Taylor, Oct. 6, 1846, Calvin Taylor and Family Collection, Department of Archives, Louisiana State University, Baton Rouge. The collection (hereinafter cited as the Calvin Taylor Collection) contains part of the Calvin Taylor Diary, all of the Sereno Taylor Diary, journals, letters, cash books, business papers, wills, land deeds, papers relating to slave purchases, and miscellaneous papers. Much of the material pertains to the Taylor sawmills located at Handsboro, Mississippi. John H. Lang, *History of Harrison County* (Gulfport 1936) 76-78, 159; *Lumber Trade Journal*, Nov. 15, 1908, 15.

[2]L. N. Dantzler Lumber Company Records, Abstract Book No. 1020, 262; Seventh Census 1850, Schedules No. 1, 5; Davis to Taylor, Oct. 6, 1846, Calvin Taylor Collection; Kremer Motors Company, land abstract records, in possession of Kremer Motors Company, Handsboro, Miss. The Kremer Motors Company was involved during the years 1930-1935 in litigation concerning land boundaries on Bayou Bernard, and possesses maps and records showing the precise location of most of the ante-bellum mills on Bayou Bernard in the vicinity of Handsboro.

[3]Seventh Census 1850, Schedule No. 5; Lang, *op. cit.* 80-81.

[4]Seventh Census 1850, Schedule No. 5; Calvin Taylor Collection, Journal No. 8, 1846-50, Taylor-Davis-Fowler Partnerships; statement made to the author by Alonzo Hickman, July 5, 1954. Hickman, a timber cruiser, estimated timber from North Carolina to east Texas in the years 1895-1945; *Mississippi Laws* (Feb. 4, 1852) 461-462. The manuscript census returns show that Brown was from Kentucky, McBean from Scotland, Fowler from New York, and Williams from Maine.

⁵*Mississippi Laws* (Mar. 8, 1852) 498-499.

⁶Calvin Taylor Collection, Charles Ludlow to Calvin Taylor, June 25, 1850.

⁷Pascagoula *Democrat Star*, Sept. 27, 1887; Calvin Taylor Collection, Sereno Taylor to Calvin Taylor, Feb. 3, 1853; *Northwestern Lumberman*, Oct. 9, 1880, 3.

⁸Wailes, Notes, 76; Seventh Census 1850, Schedules No. 1, 2, 5.

⁹Pascagoula *Democrat Star*, June 8, 1887. Wailes observed that longleaf pine was taken to the mills along the seaboard and shipped to Europe and the West Indies. Sticks of longleaf pine suitable for spars and masts were in great demand at lucrative prices; large quantities were sold to the French navy by the coast timber people. Wailes, *Agriculture*, 348-349. Another source reveals that Joseph E. Murell in 1843 contracted to furnish the French, Spanish, and British governments with square and spar timber. Several classes of spar timber ranged from fifty to eighty feet in length. The largest spar was said to have been eighty-two feet long, thirty inches at the butt end, and twenty inches at the small end. *Northwestern Lumberman*, July 18, 1885, 5; Jackson *Mississippian*, July 1, 1857.

¹⁰*De Bow's Review* XXIV (1858) 239; Jackson *Mississippian*, July 5, 1859.

¹¹Eighth Census 1860, Schedule No. 5; Lang, *op. cit.* 76, 81. The will of S. S. Henry, in the abstract records of the Dantzler Company, shows that Henry owned both gang and circular saws in 1860. L. N. Dantzler Lumber Company Records, Abstract Book No. 1084, 27. The manuscript returns show that Liddle, Taylor, and Burr were natives of New York and Dodge a native of Connecticut.

¹²Calvin Taylor Collection; Harrison County Tax Roll, 1858, 14-18, in Mississippi State Department of Archives and History, Jackson.

¹³Seventh Census 1850, Schedule No. 2; Eighth Census 1860, Schedule No. 2. Slave holdings of lumbermen, logmen, and naval stores producers in Jackson, Harrison, and Hancock counties were as follows:

Name	Number of Slaves	
	1850	1860
Bingham, H.	4	
Brown, James		35
Carr, R. W.	23	18
Champlain, W. A.		4
Dees, John	37	
Denny, Walter		33
Files, Davis	32	
Fowler, Sarah		1
Gammell, J. A.		21
Goode, Garland		39
Griffin, William		39
Hand Foundry Company		25
Hand, Miles B.	9	
Henderson, John	19	
Henry, S. S.	8	
Hester, Goodman		1
Horn, G. W.	4	1
Huddleston, John	18	24
Humphries, Thomas		30

¹⁴Calvin Taylor Collection.

CHAPTER IV

[1]Barber manuscript, 2; statement made to the author by Calvin Dees, Oct. 7, 1953. Dees's information came from his grandfather, John Dees, who was a partner of Garland Goode in the ante-bellum period; *Northwestern Lumberman*, Oct. 9, 1880, 6; Lang, *op. cit.* 73; John K. Bettersworth, *Confederate Mississippi* (Baton Rouge 1943) 183.

[2]*Northwestern Lumberman*, Nov. 6, 1880, 4; *Southern Lumberman*, Dec. 1, 1885, 9.

[3]Fries, *op. cit.* 61-62; Ninth Census 1870, Schedule No. 4; Pascagoula *Democrat Star*, May 23, 1874.

[4]Fries, *op. cit.* 61-63; *Lumber Trade Journal*, July 15, 1907, 22; Newsome Family Papers; *Southern Lumberman*, July 1, 1886, 9, July 1, 1887, 6.

[5]Ninth Census 1870, Schedule No. 4.

[6]Almost all of the material on which the remainder of the chapter is based was derived from the *Southern Lumberman*, 1881-1890; *Northwestern Lumberman*, 1875-1890; *Lumberman's Gazette*, 1873-1875; Pascagoula *Democrat Star*, 1872-1906; Sargent, *op. cit.*; Mohr, *op. cit.*; Ninth Census 1870, Schedule No. 4; Tenth Census 1880, Schedule No. 3.

[7]A more detailed discussion of depredations on timberlands owned by the Federal Government is found in Chapter VI.

[8]Sargent, *op. cit.* 531. The Federal Census of 1880 was apparently incorrect, for it listed less than ten mills from the Moss Point-Pascagoula district. Tenth Census 1880, Schedule No. 3.

[9]*Northwestern Lumberman*, Oct. 10, 1885, 16; L. N. Dantzler Lumber Company Records, Miscellaneous Papers; Barber manuscript, 6; *Southern Lumberman*, June 5, 1884, 13.

[10]Mohr, *op. cit.* 43.

[11]Maxson, *op.cit.* 12; Bettersworth, *op. cit.* 183; Ninth Census 1870, Schedule No. 4.

[12]Maxson, *op. cit.* 13.

[13]Weston Collection, Weston statement, 6.

[14]Statement made to the author by Alonzo Miles, July 1, 1954. The road was described to Miles by a number of people employed by Leinhard in 1866. Leinhard's early road was described to the author in 1936 by John Clarke who saw it in operation in 1867.

Chapter V

[1]10 *Interstate Commerce Commission Reports* (1905); *St. Louis Lumberman,* 1900-1915; *Northwestern Lumberman,* 1875-1890; *Lumberman's Gazette,* 1873-1875; *American Lumberman,* 1900-1915; *Lumber Trade Journal,* 1900-1915; *Southern Lumberman,* 1881-1915, 1931-1932. These sources are frequently drawn upon throughout this chapter.

[2]Ninth Census 1870, Schedule No. 4.

[3]10 *Interstate Commerce Commission Reports* (1905) 199.

[4]St. Louis *Press,* clipped in the Pascagoula *Democrat Star,* Aug. 1, 1874.

[5]*American Lumbermen: History* I, 369.

[6]*Timberman,* July 30, 1890, 11.

[7]*Southern Lumberman,* Aug. 15, 1888, 15. According to Mohr, *op. cit.* 43, production of lumber on the Illinois Central for the year 1879-1892 was as follows:

Year	Board Feet	Year	Board Feet
1879-1880	12,000,000	1887-1888	62,000,000
1883-1884	28,000,000	1888-1889	52,000,000
1884-1885	36,000,000	1889-1890	64,000,000
1885-1886	30,000,000	1891-1892	78,240,000
1886-1887	40,000,000	1892-1893	181,424,000

[8]Isaac Enochs, *Corporate Holdings of Timberland* (Syracuse 1906) 9.

[9]Ninth Census 1870, Schedule No. 4. According to Mohr only 12,000,000 board feet were shipped from mills located adjacent to the Mobile and Ohio Railroad in 1891-1892. Mohr, *op. cit.* 43.

[10]*American Lumbermen: History* III, 253-254, 257-259.

[11]Biloxi *Herald,* Feb. 15, 1902, Feb. 18, 1888, Mar. 6, 1888.

Chapter VI

[1]*Mississippi Senate Journal* (1850) 50; *Report of the Secretary of the Treasury* 1848, 30th Cong., 2nd Sess., Senate Exec. Doc. 2 (1848) 247.

[2]L. N. Dantzler Lumber Company Records, Tract Book No. 1, Jackson County; Stone County Archives, Tract Book A, Harrison County; *Report of the Commissioner of the General Land Office* 1851, 32nd Cong., 1st Sess., Senate

Exec. Doc. 1 (1851-1852) 23, 29; 1853, 33rd Cong., 1st Sess., Senate Exec. Doc. 1 (1853) 92, 99.

³L. N. Dantzler Lumber Company Records, Tract Book No. 1, Jackson County; Stone County Archives, Tract Book A, Harrison County.

⁴14 *U. S. Statutes at Large* (1866) 66-67.

⁵*Congressional Record*, 44th Cong., 1st Sess., Vol. 4, Pt. 1 and Pt. 4 (1876) 850-852, 817-818, 3293-3294. Senator Alcorn stated that the Federal Government owned 4,263,520 acres in Mississippi, mostly barren pineland, and that the state owned 2,000,000 acres acquired through tax sales.

⁶19 *U. S. Statutes at Large* (1877) 73-74; L. N. Dantzler Lumber Company Records, Land Ownership Journals No. 1, 2, 3. These records give names of original buyers and date of sale of all lands owned by the company.

⁷*Report of the Commissioner of the General Land Office* 1877, 16-21; Fries, *op. cit.* 185-193.

⁸The material dealing with depredations on U. S. timberlands comes chiefly from the report on the seizure of logs, timber, and naval stores, Senate Exec. Doc. 9 (1877-1878) Pt. 1, 64-65, Pt. 2, 8-9, 38-41; *Congressional Record*, 45th Cong., 2nd Sess., Vol. 7, Pt. 2, 1398-1408, 1524, 1942; *Congressional Record*, 46th Cong., 2nd Sess., Vol. 10, Pt. 4, 3627-3632. The most complete coverage was found in the Pascagoula *Democrat Star* 1877-1879. The Pascagoula paper in Nov. 30, 1877, gave the values of lumber, logs, and timber seized by the Federal authorities from individuals and firms as follows:

A. C. Danner	$ 16,750
De Smet	13,050
Denny and Morris	900
L. N. Dantzler	12,800
E. F. Griffin	33,000
Gautier and Sons	2,400
Wallworth	2,400
John McClean	2,424
W. G. O'Neal	15,000
Wyatt Griffin	37,500
Total	$136,224

According to President Rutherford B. Hayes there were forty-nine lawsuits involving $200,000 in southern Mississippi. *Report of the Commissioner of the General Land Office* 1879, 558-559. Garland Goode, owner of 40,000 acres of timberland, stated that millmen were responsible for depredations because of the low prices they paid for logs. A logman could not buy timber and sell it profitably to the millmen in competition with many others who got their logs without cost from the public domain. Pascagoula *Democrat Star*, Jan. 4, 1878. George Leatherbury, a lumberman and naval stores operator, was indicted for boxing timber on government-owned land. Attorneys argued that boxing timber was not a violation of Federal laws since timber was not cut or severed from the land. The court held that cutting timber to extract the gum and sap for one's private use was cutting with intent to use in a manner other than for the Navy of the United States. 27 *Federal Reporter* (1886) 606.

⁹Pascagoula *Democrat Star*, April 9, 1881, Feb. 9, 1886, Jan. 13, 1893.

[19]*Congressional Record*, 44th Cong., 1st Sess., Vol. 4, Pt. 1, 852; *Report of the Commissioner of the General Land Office* 1879, 609; *Northwestern Lumberman*, Oct. 9, 1880, 2, June 25, 1887, 3-6; Thomas Donaldson, *The Public Domain: Its History, with Statistics* (Washington 1884) 530; *Report of the Commissioner of the General Land Office* 1883, 9; Pascagoula *Democrat Star*, April 16, 1886; *Report of the Commissioner of the General Land Office* 1887, 505.

[11]Paul Wallace Gates, "Federal Land Policy in the South, 1866-1888," *Journal of Southern History* VI (Aug. 1940) 316-321; L. N. Dantzler Lumber Company Records, Log Book No. 2, 1912; *Lumber World Review*, Nov. 10, 1912, 44; *St. Louis Lumberman*, Nov. 15, 1912, 45; *Timberman*, Aug. 18, 1890, 11; *American Lumbermen: History* I, 235-238. The following table is from Gates:

SOUTHERNERS

Name	Residence	Acres
Butterfield, J. S.	Brookhaven, Mississippi	20,599
Brackenridge, E. A.	Orleans Parish, Louisiana	7,520
Colley & Warner	Marion County, Mississippi	8,580
W. F. Evans & Co.	Lauderdale, Mississippi	6,323
Griffin & Perkins	Perry County, Mississippi	10,406
Kemper, J.	Laurel, Mississippi	14,600
Leinhard, H.	Harrison County, Mississippi	7,524
Orrell, J. C., *et al.*	Jackson County, Mississippi	17,068
Persons, J. W.	Lincoln, Mississippi	5,580
Richardson, W. P.	Hinds County, Mississippi	21,210
Waddell, S.	Union City, Tennessee	14,860
Total		134,270

NORTHERNERS

Name	Residence	Acres
Bewick & Comstock	Detroit, Michigan	65,486
Blodgett, D. A.	Grand Rapids, Michigan	136,238
Birkett, McPherson, *et al.*	Howell, Michigan	18,746
Cartier & Dempsey	Manistee, Michigan	9,563
Chesbrough, A. M.	Toledo, Ohio	13,180
Conkling, O. F.	Grand Rapids, Michigan	9,120
Doud & Bonner	Winona, Minnesota	6,202
Hamlin, H.	Smithport, Pennsylvania	12,195
Heald & Nufer	Montague, Michigan	15,377
Henry, F.	Warren County, Pennsylvania	9,600
Hills, C. T.	Muskegon, Michigan	69,828
Kent, G.	Delhi, Ontario	6,169
McKeown, J.	Parker City, Pennsylvania	24,646
Moores & McPherson	Lansing, Michigan	11,425
Plock, O.	New York, New York	42,588
Randsdell, D. M.	Indianapolis, Indiana	69,843
Rich, S. B.	Wayne County, Michigan	18,052
Robinson & Avery	Detroit, Michigan	22,485
Robinson, Lacey, *et al.*	Grand Rapids, Michigan	6,360
Sage, H. W.	Ithaca, New York	34,559
Schlesinger, B.	Boston, Massachusetts	7,475
Southwell, H. E.	Milwaukee, Wisconsin	5,420
Squier, D. W., A. T., & F. W.	Ashland, Michigan	39,648

Tippin, G.	Defiance, Ohio	9,672
Tomlinson, S. J.	Lapeer County, Michigan	5,960
Tuttle, B. B.	Naugatuck County, Connecticut	10,015
Vaughan & Johnson	Lapeer County, Michigan	8,509
Wagar & Wells	Ionia, Michigan	17,680
Ware & Blanchard	Grand Rapids, Michigan	10,090
Watson, A. B.	Grand Rapids, Michigan	82,885
Weston, I. M.	Grand Rapids, Michigan	33,716
Wilson, R. T.	New York, New York	49,717
Total		889,359

[12]*Congressional Record*, 50th Cong., 1st Sess., Vol. 19, Pt. 4 (1888) 3032; Biloxi *Herald*, May 10, 1888; 25 *U. S. Statutes at Large* (1888) 622.

[13]*Report of the Commissioner of the General Land Office* 1890, 109-110, 252; Mississippi State Land Office, Patent Book No. 2.

[14]Donaldson, *op. cit.* 1056; statement of Billy Batson; *Report of the Commissioner of the General Land Office* 1877, 24-25; Stone County Archives, Tract Books A and B; L. N. Dantzler Lumber Company Records, Tract Books No. 1, 2; Enochs, *op. cit.* 6-8.

[15]Statement of Billy Batson, July 6, 1954. Batson, a timber cruiser, was familiar with the methods of homesteading in the timber country. Statements of Griffin and John Hickman.

[16]Pascagoula *Democrat Star*, Jan. 8, 1889, Aug. 6, 1900; statements of Griffin and Batson.

Chapter VII

[1]*Report of the Commissioner of the General Land Office* 1898, 267.

[2]Mississippi State Land Office, Records of the Augusta Land District, Tract Books No. 1, 2, 3; Stone County Archives, Tract Book A; L. N. Dantzler Lumber Company Records, Jackson County Tract Book No. 1; *Report of the Commissioner of State Lands of the State of Mississippi* 1896-1897, 170-182; Pascagoula *Democrat Star*, Aug. 6, 1874; *Mississippi House Journal* (1898) 12, 15-16; Donaldson, *op. cit.* 228-230. The Mobile and Ohio Railroad was granted 1,004,640 acres in Mississippi, but the amount actually obtained was estimated at 737,130 acres. A large portion of the Mobile and Ohio lands was located in the longleaf pine counties of Greene, Wayne, and Clarke. The original grant to the Gulf and Ship Island Railroad in 1856 was 656,800 acres. The amount actually received by the railroad was much less. *Report of the Commissioner of the General Land Office* 1875, 404.

[3]9 *U. S. Statutes at Large* (1850) 519-520.

[4]*Report of the Commissioner of Swamplands of the State of Mississippi* 1877, 4.

[5]133 *Mississippi Reports* (1923) 373; *Mississippi Laws* (March 16, 1852) 31-32.

[6]*Report of the Commissioner of State Lands* 1892-1893, 10; Stone County Archives, Tract Book B. Most of the lands cleared by farmers in northern Harrison County in the fifties were so-called swamplands selected by the state. Large tracts consisted of high, dry pinelands. L. N. Dantzler Lumber Company Records, Tract Book No. 2, Jackson County; statement made to the author July 16, 1953, by Pete McLeod, a lawyer of Jackson County. 33 *U. S. Statutes at Large* (1905) 1258; *Mississippi Laws* (Feb. 22, 1890) 303-304.

[7]*Mississippi Laws* (1852-1890); *Mississippi Reports* (1923); *Reports of the Land Commissioners* 1882 through 1885; Stone County Archives; L. N. Dantzler Lumber Company Records; Gates in *Journal of Southern History* VI, 325; 11 *U. S. Statutes at Large* (1856); *Mississippi Senate Journal* (1882); *Mississippi House Journal* (1886).

[8]L. N. Dantzler Lumber Company Records, Abstracts. The company purchased about 20,000 acres of timberland from Rankin Hickman, who had acquired his holdings from the state in entering land in 240-acre tracts through other individuals. According to Wilmer Griffin, the Griffins acquired a large body of timberland from the state in similar fashion, 240-acre blocks having been purchased even in the names of their oxen.

[9]*Mississippi House Journal* (1886) 148, 555, 570; L. N. Dantzler Lumber Company Records, Abstract Book 1020.

[10]*Mississippi Laws* (1890) 30-31, (Feb. 17, 1890) 34, (1892) 279-282; *Biennial Report of the Commissioner of State Lands* 1892-1893.

[11]Statements of John Hickman and Batson.

[12]*Biennial Report of the Commissioner of State Lands* 1896-1897, 5, 158.

[13]*Mississippi Senate Journal* (Special session, 1898) 10. An act of 1818 had provided for leasing of sixteenth-section lands for periods of three years; an act of 1824 extended the time to five years; an act of 1833 authorized a ninety-nine year lease. 42 *Southern Reporter* (1907) 290-291.

[14]Statement of Daniel McQuagge. The board of supervisors of the county was offered $1,800 for a turpentine (naval stores) lease on a sixteenth section of land, but refused. Shortly afterwards the supervisors leased the land for $50. In another instance in the same county, the supervisors leased a section of virgin timberland for $20.

[15]42 *Southern Reporter* (1907) 300-301; Stone County Archives, Minutes of the Board of Supervisors 1917, Book I. The Mississippi Code of 1906 authorized the sale of marketable timber on sixteenth sections by county boards of supervisors. The proceeds of such sales were to be loaned out by county officials and the interest therefrom used in support of common schools. 53 *Southern Reporter* (1910) 2.

[16]Pascagoula *Democrat Star*, Jan. 4, 1910, Jan. 4, 1901; *Mississippi House Journal* (Special session, 1898) 15-16; Stone County Archives, Tract Book A; L. N. Dantzler Lumber Company Records, Jackson County Tract Book No. 1; Mississippi State Land Office, Patent Book No. 2; Gates in *Journal of Southern History* VI, 326.

[17]*Report of the Commissioner of the General Land Office* 1875, 404; Pascagoula *Democrat Star*, Aug. 6, 1874; Enochs, *op. cit.* 6-7.

[18]Mississippi State Land Office Patent Book No. 2, 2-50; L. N. Dantzler Lumber Company Records, Abstract Book 1470, 5; *Report of the Commissioner of State Lands* 1895-1896, 9.

Chapter VIII

[1]Statement of Griffin.

[2]Calvin Taylor Collection, Journal No. 9, 1856-1862, 20-40; statement of Miles. Miles cut logs with an ax for Henry Leinhard in the late eighties. Statement of John F. Hickman. Hickman cut cypress for a short time in the forks of Red and Black creeks in 1890. The technique of chopping cypress differed slightly from that used in chopping pine. Most of the cypress grew in lakes or ponds, and the base of the tree, from five to seven feet above the ground, was not uniformly rounded. The choppers built a scaffold upon which they could stand high enough to cut the tree above its uneven surface. This operation often resulted in a cypress stump four to seven feet tall. Long before the actual felling of cypress began, the trees were girdled so that they would die and dry out; this was essential because the heavy green cypress logs would not float. Cypress, considered unusually valuable at that time, was perhaps the first timber, with the possible exception of spars, to be rafted on the streams of south Mississippi. Wailes, *Agriculture*, 348-349. An anonymous writer stated in the Pascagoula *Democrat Star*, June 8, 1877, that spar timber had been shipped from the mouth of the Pascagoula for thirty-five years.

[3]Statement of Griffin.

[4]*Northwestern Lumber*, July 16, 1887, 18; *Lumber Trade Journal*, July 15, 1903, 24; statement of Walter Barber.

[5]Observations of the author. Stringers and sills of one dwelling still in use were said to have been cut by the author's grandparent in 1870. Shipment of ties to overseas markets from Mobile was of considerable importance. The American Railway Engineering and Maintenance of Way Association recommended that ties be eight by eight inches, ten feet in length. *Lumber Trade Journal*, April 1, 1902, 22, 29.

[6]Statements of Griffin, John Hickman, and Barber.

[7]Calvin Taylor Collection; Wailes, *Agriculture*, 348-349; Maxson, *op. cit.* 11; statements of Griffin and John Hickman; Cain, *op. cit.* 145. All parts of the caralog except its iron treads were manufactured in pioneer blacksmith shops scattered throughout the piney woods. The frame of the draketail caralog was constructed on one pine pole, the small end extended out in front to form the tongue, and the other end, split into two sections, extended to the rear where a windlass was attached to hoist the log. The framework was balanced on the axle of the two-wheeled cart, and grab hooks dropped down to encircle the timber. The windlass hoisted one end of the log to the axle to be fastened, while the other was allowed to rest on the ground. The maximum carrying capacity of the cart was limited to three small logs, or two large ones.

[8]George H. Dacy, *Four Centuries of Florida Ranching*, as quoted in Cain, *op. cit.*, 129.

[9]Observations of the author in Stone County during the years 1920-1930; *Southern Lumberman*, Dec. 15, 1856, 134; statements of Dees and John Hickman; Cain, *op. cit.* 129-130; Agnes M. Larson, *History of the White Pine Industry in Minnesota* (Minneapolis 1949) 78; Pascagoula *Democrat Star*, July 8, 1882. The writer stated that some of the best drivers on Red Creek got from $1.00 to $2.50 per day. Babe Fairley, who for more than thirty years drove ox teams for many different people, stated to the author Nov. 1,

1953, that wages of ox drivers in the years 1885-1920 varied from seventy-five cents per day to $2.00; statements of John Hickman and Miles.

[10]L. N. Dantzler Lumber Company Records, Moss Point Lumber Company Log Books, Vols. 1, 2, 3, 4, 5, 6, 1892-1901. The log tallies show that one logman might have logs bearing as many as twenty different brands. L. N. Dantzler Lumber Company Records, Log Brand Book. This book contains a list of over 5,000 brands used on the Pascagoula and its tributaries in the years 1865-1930; Sixth Census 1840, *Compendium*, 236; Seventh Census 1850, Schedule No. 5; Wailes, Notes, 34; Wailes, *Agriculture*, 348-349; Calvin Taylor Collection, Journal No. 9, 1858-1862, 5-200; L. N. Dantzler Lumber Company Records, Miscellaneous Papers. Some of the papers concern the business affairs of S. S. Henry and show that twenty-five logmen in Harrison County were selling logs to him in 1858-1860; Hilgard, *op. cit.* 382-385; *Mississippi Laws* (Feb. 10, 1860) 354-355; statement of John Hickman. Hickman's father, Rankin Hickman, who began rafting on Red Creek and the Pascagoula River in 1867, stated that the ditch connecting Black and Red creeks was dug long before his first trip down to Moss Point. Although the details of logging and rafting in the ante-bellum period are rather obscure, their importance to the economy was attested by an act of the state legislature in 1860. This act empowered the board of police in Harrison County to allocate proceeds derived from swampland sales to clearing, straightening, and widening streams. Such improvement of navigation would expedite the floating of log rafts and movement of spar timber. Another project at public expense that testified to the importance of log-running was the diversion of Black Creek, which originally flowed directly into the Pascagoula River, into Red Creek and Dead Lake. By this change all the timber coming down Black Creek could be boomed at Dead Lake preparatory to being sent down the Pascagoula to tidewater.

[11]Statements April 6, 1953, by Adam Blackwell, who rafted logs on the Biloxi rivers in the eighties, and of Miles who also rafted on the Biloxi rivers, and those of John Hickman, McQuagge, and Fairley; Bond, *op. cit.* 8; Cain, *op. cit.* 146-147.

[12]For this reason men were stationed at sharp bends along the river, called check points, to assist the crewmen on the raft. When the floating timber approached the check point, a rope was tossed from the raftsman to a man on shore who tied it to a tree. If the maneuver was successful, the pressure exerted by the rope caused the raft to swing clear of the land. If such efforts of the crewmen were unsuccessful, the logs traveling at a rapid pace ran headlong into the shore and broke up. Pascagoula *Democrat Star*, June 4, 1874; *American Lumberman*, April 28, 1900, 38; statement of John Hickman, who helped break the jam and reported that more than forty days elapsed before all the stranded logs were moved down the river.

[13]Bostick Breland, unpublished manuscript, 125. Victor Scanlon who operated a log business on the Leaf River and later built a big mill at Clyde, Mississippi, stated that the two happiest years of his life were spent in the wilderness with the rough, hardy folk who were part-time farmers and rafters. According to Scanlon, the piney woods people were not daunted by weather or physical discomfort. He was particularly impressed by the life of indescribable toil and hardship endured by the woods folk. Some of the people floated their logs down to Moss Point and walked back home, a distance, in

some cases, of more than one hundred miles. Hattiesburg (Miss.) *American*, Mar. 6, 1957.

[14]Statements of John Hickman, Rankin Hickman, and Fairley.

[15]L. N. Dantzler Lumber Company Records, Moss Point Lumber Company Log Books, Vols. 1, 2, 3, 1892-1896; statements of John Hickman and Ward.

[16]*Mississippi Laws* (Feb. 27, 1886) 174-176; Moss Point (Miss.) *Journal*, March 26, 1888.

[17]Pascagoula *Democrat Star*, June 19, 1889, April 14, 1893.

[18]L. N. Dantzler Lumber Company Records; Moss Point Lumber Company Records, Log Books, Vols. 1, 2, 4, 1892-1896; statements of John Hickman and Roy Baxter, July 8, 1953, the latter an administrative officer of the H. Weston Company of Logtown, Miss.

[19]Statements of Fairley and John Hickman. Randolph Batson began as a log chopper and eventually became one of the larger yellow pine manufacturers in Mississippi.

[20]L. N. Dantzler Lumber Company Records, Miscellaneous Papers; Moss Point Lumber Company Records, Log Books, Vol. 3, 1892-1915, 22-26; Pascagoula *Democrat Star*, May 4, 1896; statement of Barber.

CHAPTER IX

[1]Resin in common parlance was another name for crude gum. Rosin was one of the products of naval stores production. Carl E. Ostrom, "History of the Gum Naval Stores Industry," *The Chemurgic Digest* IV, No. 12 (July 15, 1945) 217, 220-221; Thomas Gamble, *Naval Stores: History, Production, Distribution, and Consumption* (Savannah 1921) 127; statement of Simpson.

[2]Statement made July 4, 1954, by John Gary, then in his late seventies, who began chipping turpentine trees in Georgia at the age of eighteen; statement of Wes Tatum, Oct. 6, 1953. Tatum, who operated a fairly large sawmill enterprise, stated that because of the damage to timber, his company did not engage in naval stores operation. According to Simpson, the September storm of 1906 blew down most of the turpentined timber in southeastern Mississippi, thereby putting many of the small naval stores operators out of business.

[3]Statements of Gary and Simpson. The chipping tool known as the hack evolved over a long period. The prototype of the hack was called the round shave in the 1840's, and consisted of a steel blade shaped like an individual's forefinger curved from the second joint. About mid-century the tool which became known as the hack began to be manufactured. The size of the hack blades governed the depth and width of the streak. In the early days the streaks were ordinarily one inch in depth and width; subsequently it was learned that a shallower and narrower wound increased the flow of resin. Ostrom, in *The Chemurgic Digest*, 220.

[4]Statements of Gary and Simpson.

[5]Gamble, *op. cit.* 25, 27, 222; statements of Gary, Simpson, and Leonard O'Neill, July 15, 1954; statement of Dennis Smith, July 12, 1954; statement of

Peter Fairley, July 20, 1954. O'Neill at the age of eighteen began racking boxes and dipping turpentine in Georgia. Later he came to Mississippi and was employed by various operators as a woods-rider. For a number of years he was superintendent of the naval stores business of the L. N. Dantzler Lumber Company. Smith, born in Georgia, began chipping boxes at the age of eighteen. He has chipped trees in every state from Georgia to east Texas over a sixty-year period. At the age of eighty Smith still continued his lifelong occupation. Fairley, who is seventy years old, has chipped boxes since 1904 in Mississippi, Louisiana, and east Texas.

⁶Ostrom, op. cit. 227; Mohr, op. cit. 68-69; A. W. Schorger and W. S. Betts, *The Naval Stores Industry*, U. S. Forest Service Bull. No. 229 (Washington 1945) 2-3.

⁷Mohr, op. cit. 69; Sixth Census 1840, *Compendium*, 230; Raymond *South-Western Farmer*, Jan. 20, 1845, 155.

⁸Gainesville (Miss.) *Advocate*, May 8, 1846.

⁹Pascagoula *Democrat Star*, May 8, June 22, 1889; Wailes, *Agriculture*, 350-351.

¹⁰Seventh Census 1850, Schedule No. 5; Cain, op. cit. 148-149.

¹¹Seventh Census 1850, Schedule No. 5; Ninth Census 1870, Schedule No. 4; Sargent, op. cit., map opposite 530. The Pascagoula *Democrat Star*, April 24, 1877, estimated that annual production of turpentine in Mississippi was 50,000 barrels; of resin, 250,000 barrels, with the total value of naval stores $160,000 annually.

¹²Sargent, loc. cit.; Pascagoula *Democrat Star*, July 9, 1886; Tenth Census 1880, II *Report on the Manufactures of the United States*, 141; Gamble, op. cit., 78, 109-110; L. N. Dantzler Lumber Company Records; statement of John Hickman who in the late eighties and early nineties hauled freight from Leatherbury's commissary located at Benndale on the west bank of the Pascagoula.

¹³Enochs, op. cit. 6; *Report on the Internal Commerce of the United States*, 49th Cong., 2nd Sess., House Exec. Doc. 7 (Dec. 20, 1886) Pt. 2, 502; Sargent, op. cit. 530. Dr. Charles Mohr, who made the investigation in Mississippi in 1880, reported thirty-three stills on the line of the Mobile and Ohio Railroad between Mobile station and Quitman, Mississippi.

¹⁴Eleventh Census 1890, IX *Manufacturing Industries*, Pt. 3, 644-645; Gamble, op. cit. 78; Mohr, op. cit. 43; statement of Miles. Miles said that turpentine distilleries were erected shortly after the construction of the Gulf and Ship Island Railroad. The account given by Miles agrees with the statements of others who traveled over the railroad during the late nineties.

¹⁵Statements of Simpson, Gary, and John Hickman.

¹⁶Statement of Gary; Gamble, op. cit. 110-111; Pascagoula *Democrat Star*, Feb. 16, 1900.

¹⁷Lowe, op. cit. 133.

¹⁸Statements of Simpson and Gary; Ostrom in *The Chemurgic Digest*, 220.

¹⁹Schorger and Betts, op. cit. 19; statements of Barber and Simpson; statement of Mayers Dantzler, June 4, 1954.

²⁰*Lumber Trade Journal*, Dec. 1, 1903, 23; statements of Gary, O'Neill, Barber, and Simpson; statement of Myrrah Riley, July 20, 1954; L. N. Dantzler

Lumber Company Records, Elen Dee Naval Stores Accounts. The Elen Dee
Naval Stores Company was a L. N. Dantzler Lumber Company enterprise.
Lumber World Review, Oct. 10, 1915, 37; Gamble, *op. cit.* 111.

²¹Gamble, *op. cit.* 83-84; Lowe, *op. cit.* 134-135.

²²*Southern Lumberman,* Jan. 21, 1909, 40; *Lumber Trade Journal,* April 1,
1910, 23; *St. Louis Lumberman,* Oct. 15, 1915, 50.

²³Gamble, *op. cit.* 59-65, 71; *Lumber Trade Journal,* July 15, 1914, 16,
April 1, 1915, 22-23, June 1, 1915, 31-32, July 1, 1915, 31-32; *Southern Lumberman,* Dec. 18, 1915, 121.

Chapter X

¹Statements of Barber and Gary; statement made by Lee Taylor, July 20,
1954. Taylor, in his mid-seventies, started chipping boxes in Mississippi at the
age of eighteen. In the years 1900-1930 he was employed as a chipper in
Mississippi, Louisiana, and Texas. Both Negroes and whites are in substantial
agreement as to the character of the turpentine worker.

²Statement of Gary.

³Statement of Simpson.

⁴Statements of Peter Fairley and Smith.

⁵Statements of Simpson and Gary. Most of the states producing naval
stores enacted laws to protect operators (landlords) who made advances to
their Negro workmen. In Gamble, *op. cit.* 101-104, Albert Pridgen, who was
connected with the naval stores industry in Georgia, Alabama, and Mississippi,
stated that the system of credit was a curse. The landlord, feeling somewhat
protected by the laws, made liberal allowances which resulted in a big debt
that the Negro could never work out. The debt was held against the Negro
to keep some other operator from moving him away.

⁶Statements of Simpson, Gary, Smith, and Taylor. According to Pridgen,
the lawlessness of the Negro which characterized the majority of Negro labor
made the life of the operator one of danger. However, he believed the
turpentine Negro to be no worse than the sawmill Negro. The turpentine
camp was no place for the timorous; a man had to be ready to use force or
weapons in self-defense. Gamble, *op. cit.* 101-104.

⁷Statements of Smith, Gary, and Simpson. According to Pridgen, the pioneers
who moved west fifty years ago in the eighties and nineties did not find conditions more difficult than the modern turpentine operator who moved to a
new country. When the operator moved out into the piney woods, he almost
severed his connections with the outside world. His food was plain and he
enjoyed few if any conveniences. His company was restricted to the Negroes
under his charge, and in this environment he lost the veneer of refinement
that came from association with those of his own kind. Gamble, *op. cit.* 104.

⁸Statements of Simpson, Gary, and Smith.

⁹P. L. Buttrick, "Commercial Uses of the Longleaf Pine," *American Forestry*
XXI, 901.

¹⁰Statements of Simpson, O'Neill, Gary, Smith, and Riley.

CHAPTER XI

[1]*Lumber Trade Journal*, Mar. 15, 1902, 23.

[2]*Southern Lumberman*, Aug. 1, 1900, 8.

[3]Tenth Census 1880, II *Manufactures,* 141; Eleventh Census 1890, *Manufactures,* 482; Thirteenth Census 1910, X *Manufactures,* 505; *Census of Manufactures: 1914* (2 vols. Washington 1919) I, 761; Lowe, *op. cit.* 133; *Lumber Trade Journal,* Jan. 1, 1907, 26-27.

[4]*St. Louis Lumberman,* 1900-1915; *Southern Lumberman,* 1880-1932; *Lumber Trade Journal,* 1900-1916; *Lumber Review,* 1906-1912; *Lumber World Review,* 1912-1916; and *Northwestern Lumberman,* 1875-1890, are the chief sources for material in Chapter XI. Southern Pine Association *Statistical Report,* 1952.

[5]Biloxi *Herald, Twentieth Anniversary Edition,* 72-73; 10 *Interstate Commerce Commission Reports* (1905) 517.

[6]*Tariff Hearings* 1908-1909, U. S. House Committee on Ways and Means, 60th Cong., 2nd Sess., House Doc. 1505 (9 vols., Washington 1909) III, 2893. This document is hereinafter cited as *Tariff Hearings.* Lowe, *op. cit.* 49.

[7]Statement of Barber. According to one writer either a band saw and a gang resaw, or a two-band mill, usually proved to be the best equipment for sawing yellow pine. When a larger capacity was demanded, two band saws and a gang resaw gave satisfactory results. *St. Louis Lumberman,* Feb. 15, 1912, 73. The origin of the band saw is obscure. In 1808 William Newberry of London was granted a patent on such a saw. Around 1850 J. L. Perrin of Paris received the Grand Cross of the Legion of Honor for his improvements upon the saw. *Northwestern Lumberman,* Feb. 1875, 33-34, Nov. 1875, 150-151. D. Clint Prescott was credited with the invention in the sixties of shotgun feed, a device said to have added one-third to the capacity of the circular and band saws. Prescott's invention inspired many other mechanical devices for handling and sawing timber in the mills such as steam niggers, log jumpers, loaders, and transfers. Balanced valves produced by Prescott gave sawyers almost perfect control over sawing operations. Gang slab and edging slashers were said to have been devised also by Prescott and first used in the Kirby Carpenter Company mills at Menominee, Michigan. *Lumber Trade Journal,* Feb. 1, 1903, 14. In the United States one of the first to make use of the band saw for sawing logs was Jacob R. Hoffman. In 1875 the Hoffman saw was guaranteed to cut 25,000 board feet daily. *Northwestern Lumberman,* Feb. 1875, 33-34. The band saw developed by the Hoffmans sawed a one-twelfth inch kerf. The circular saw, cutting a one-fourth inch kerf, converted the equivalent of 6,250 board feet of lumber into waste for every 25,000 board feet manufactured. In Mississippi the Norwood and Butterfield Lumber Company was credited with installing one of the first, if not the first, band sawmills in 1888 or 1890. *Lumber Trade Journal,* Mar. 15, 1910, 11; *Northwestern Lumberman,* Nov. 1, 1890, 13; *St. Louis Lumberman,* Mar. 15, 1905, 56; *American Lumberman,* April 20, 1907, 16.

[8]Statement of Barber; observations of the author in the 1920's.

[9]Statements of J. Roland Weston, Barber, and Mayers Dantzler. Statement made Sept. 1, 1954, by Joe Broadus, who was employed by the Finkbine Lumber Company as a locomotive engineer in the years 1915-1930.

[10]Statement of Clarence Hinton. Mechanization of logging equipment was fairly general and types of skidding machines and log loaders showed considerable variation. In 1910 the Russell Wheel and Foundry Company of Detroit brought out a skidder-loader combination machine for the Ingram-Day Lumber Company at Lyman, Miss. It was a straddling type, raised and lowered by hydraulic jacks. This arrangement allowed flat cars to pass under the machine. Self-propelled loader-skidders moved under their own power from one flat car to another. George Blackledge, logging superintendent of the Gilchrist-Fordney Lumber Company, stated that his company built six logging spurs to the mile. Clyde type skidders were set up near the spur and snaked logs from a distance of 800 feet on both sides of the tracks. Three ox teams were also used to snake the scattered timber. The Newman Lumber Company, the Dantzler Lumber Company, and the H. Weston Company used oxen and Lindsey wagons to bring timber to logging spurs. Statements of Barber and Weston. The Lindsey eight-wheel wagon, built by John Lindsey and Dr. S. W. Lindsey at Laurel, Mississippi, and designed especially to carry heavy loads of logs, had eight points of contact with the road and was capable of hauling 20,000 pounds at one time. As late as 1928 the Lindsey wagon was still being used in a few logging operations. Observations of the author in Stone, Pearl River, and Perry counties, 1928.

CHAPTER XII

[1]Statement of Barber; Pascagoula *Democrat Star*, Jan. 22, 1897.

[2]*Lumber Trade Journal*, April 1, 1901, August 15, 1901, August 1, 1913; *St. Louis Lumberman*, Dec. 15, 1912; *American Lumberman*, Mar. 10, 1906, April 20, 1906; statements of Weston and Dantzler.

[3]L. N. Dantzler Lumber Company Records; Barber manuscript, 3, 6-8; statements of Barber, Dantzler, and John Hickman.

[4]Almost all of the material in the remainder of the chapter was derived from the following sources: *American Lumberman*, 1902-1906; *Lumber Review*, 1906-1908; *Lumber Trade Journal*, 1901-1915; *Lumber World Review*, 1911-1915; *St. Louis Lumberman*, 1902-1914; *Southern Lumberman*, 1892-1913; *American Lumbermen: History* I, 368-369, II, 261-264.

[5]*St. Louis Lumberman*, April 15, 1905. Six planing mills at Hattiesburg had an annual capacity of 300,000,000 board feet. Owners of planing mills in 1905 were J. J. Newman Lumber Company, B. T. Toomer and Company, Brookhaven Lumber Company, Union Lumber and Planing Company, Brown Lumber Company, and J. A. Foy and Company.

[6]Statements of George Guild, Simpson, and Mrs. Finkbine.

[7]10 *Interstate Commerce Commission Reports* (1905) 516-517, 521.

[8]Larson, *op.cit.* 380.

CHAPTER XIII

[1]Shipment of lumber from Gulfport in the years 1902-1915:

Year	Board Feet	Year	Board Feet
1902	19,035,000	1909	251,492,000
1903	105,849,252	1910	300,713,000
1904	245,213,829	1911	379,932,000
1905	207,614,000	1912	315,976,000
1906	293,125,000	1913	320,816,000
1907	286,565,000	1914	182,180,000
1908	218,732,000	1915	105,030,000

These figures were tabulated from lumber journals previously cited.

[2]The trade journals previously mentioned are frequently drawn upon in Chapter XIII.

[3]*Lumber Trade Journal*, Jan. 1, 1901, 11. In some cases millmen organized their own individual firms; in others a number of export manufacturers went together to establish an overseas marketing organization. The Standard Export Company, which conducted the largest export business on the Mississippi Coast in the years 1912-1915, was an organization owned by the L. N. Dantzler Lumber Company and Price and Pierce, international timber merchants. *Southern Lumberman*, Jan. 2, 1915, 26; *Lumber World Review*, July 10, 1914, 37; *Lumber Trade Journal*, Feb. 15, 1904, 23-24.

[4]*Lumber Trade Journal*, June 1, 1904, 28, Jan. 15, 1909, 12. One reason for the formation of a manufacturer's association was the need to develop an efficient method of dealing with reclamations made by foreign purchasers of American lumber. The association from time to time maintained paid agents in foreign countries to represent members of the association in the adjustment of claims. *Lumber Trade Journal*, Jan. 15, 1909, 12, Sept. 1, 1910, 17; *Southern Lumberman*, July 27, 1907, 24, Mar. 19, 1910, 43.

[5]*Southern Lumberman*, Mar. 15, 1893, 11, 19. The association was composed of millmen east of the Mississippi River. George Robinson, L. N. Dantzler, J. A. Favre, A. S. Denny, and J. S. Otis, Mississippi coast export-producers, were members of the association.

[6]*Southern Lumberman*, June 1, 1899, 6; *American Lumberman*, Feb. 15, 1900, 13, April 14, 1900, 13, Sept. 1, 1900, 29. The company was given power to regulate output and to fix prices for all firms owning stock. Daily estimated output of the company was 1,210,000 board feet, about one-half provided by five Mississippi export-producers. Poitevent and Favre and the L. N. Dantzler Lumber Company had no connection with the firm. The company was actually a selling agency.

[7]*American Lumberman*, April 22, 1902, 33. At the time the association was formed in 1899, Cuban lumber sold at the mills for $8.50 per thousand board feet; two months later it had risen to $10.50. Prime lumber went from $18 to $20 per thousand board feet shortly after the association came into existence. Within a few months kiln dried saps advanced first from $9 to $12, and then to $15. The influence of the association in bringing about the sudden price rise is a moot question. Prices of domestic lumber advanced during the period in about the same degree, and it is probable that the change in price level was due to general economic conditions. *Southern Lumberman*, Sept. 15, 1900, 6-7.

[8]*Lumber Trade Journal*, May 15, 1906, 38, June 15, 1906, 32-34, July 1,

1906, 35, July 15, 1906, 31. During the first half of 1906 prices attained an average on a higher level than previously known. Sawn timber of thirty-five cubic feet in volume sold at 31½ and 30½ cents; that of twenty-five cubic feet, for 28 cents; that of twenty cubic feet, for 26 cents; Del Rio Plate lumber, at $20 to $21 per thousand feet; prime lumber, at $32 per thousand board feet; flooring (presumably the best grade) at $40 per thousand board feet at shipside. *St. Louis Lumberman*, Mar. 1, 1906, 41, April 15, 1906, 38. Output of sawn timber in early 1906 was low presumably because of a growing shortage of suitable timber. Timber sawn from shortleaf pine was not favorably regarded on the foreign market. In February 1906, sawn timber prices at Mobile were 27½ cents per cubic foot, the highest price recorded up to that time. *St. Louis Lumberman*, Feb. 1, 1906, 51. In May of 1906 the supply of sawn timber at Pascagoula was believed to be the shortest in the history of the trade. The L. N. Dantzler Lumber Company sold 700 pieces averaging 31½ cubic feet for 30 and one-third cents per cubic foot. *Lumber Trade Journal*, May 15, 1906, 38. In September 1906, the supply of export stock was low and every mill was busy. *American Lumberman*, Sept. 15, 1906, 57.

⁹*Lumber Trade Journal*, Aug. 1, 1907, 18-19. Exporters and manufacturers from Mobile, Pensacola, Pascagoula, Biloxi, Gulfport, and New Orleans sought to develop a uniform grade classification contract and an equitable method of settling claims. Gulf Coast exporters and manufacturers lost an estimated $1,500,000 annually through reclamations. *St. Louis Lumberman*, Aug. 1, 1907, 60.

¹⁰*St. Louis Lumberman*, July 15, 1913, 44; *Lumber World Review*, July 1913, 43, Jan. 25, 1914, 69, Feb. 25, 1914, 62. The opening of World War I in 1914 brought at first an almost complete cessation of export business. Lumber at shipside awaiting shipment was returned to the mills, and vessels loaded for overseas discharged their cargoes. Prices on the export market dropped from 20 to 25 per cent below domestic prices. Export mills curtailed production because of the scarcity of orders from England and other countries. *Lumber World Review*, Aug. 25, 1914, 45; *Lumber Trade Journal*, Aug. 15, 1914, 30, Nov. 1, 1914, 11; *St. Louis Lumberman*, Sept. 15, 1914, 44; *Southern Lumberman*, July 24, 1915, 30.

¹¹*Lumber Trade Journal*, Nov. 1, 1914, 11. Trade between American manufacturers and Latin America was facilitated by German and English-owned banking firms, with transactions based as a rule upon British sterling and the Latin American buyers generally meeting their obligations by drafts upon a British bank located in England or the United States. *Lumberman's Review*, Oct. 1914, 15. Disruption of foreign exchange compelled the South American countries to limit their imports to the bare necessities. The first nine months of the war brought a decrease of all overseas lumber exports from the Gulf ports amounting to 199,000,000 board feet. *Lumberman's Review*, Nov. 1915, 18.

CHAPTER XIV

¹10 *Interstate Commerce Commission Reports* (1905) 517; *Lumber Trade Journal*, Feb. 1, 1904, 22.

[2]U. S. Bureau of Corporations, *The Lumber Industry* IV, 75-80; the main sources upon which the narrative is based are the lumber journals.

[3]U. S. Bureau of Corporations, *op. cit.* 79-91.

[4]*American Lumberman*, Nov. 23, 1907, 37-39. In November, F. L. Peck closed one big mill and reduced operating time in another. The J. J. White Lumber Company had no orders. Most of the mills between Hattiesburg and Jackson, Mississippi, on the Gulf and Ship Island Railroad were idle, according to the *St. Louis Lumberman*, Nov. 15, 1907, 75.

[5]*Lumber World Review*, Oct. 4, 1911, 25, Jan. 10, 1913, 23. The bonds were bought by the Interstate Trust and Banking Company, and the Continental Trust and Savings Bank. The Interstate Trust and Banking Company, capitalized in 1902 at $1,500,000, was a New Orleans bank owned by lumbermen. Wallace B. Rogers, affiliated with the Eastman-Gardiner Lumber Company at Laurel, Mississippi, was president. Stockholders included I. C. Enochs, W. T. Joyce, Long-Bell Lumber Company, Ruddock Cypress Company, Frederick Weyerhaeuser, and J. J. White. *St. Louis Lumberman*, July 1902, 34.

[6]U. S. Bureau of Corporations, *op. cit.* 89-93; *St. Louis Lumberman*, Jan. 1, 1912, 44, 46, April 15, 1912, 44, 47, July 1, 1912, 44, 96, Dec. 15, 1912, 46, June 1, 1913, 48; *Lumber World Review*, April 10, 1912, 49, June 25, 1912, 50, Nov. 10, 1912, 49-50. F. A. Farwell, a lumber journalist, stated that in the years 1907-1911 most lumbermen operated their mills for returns that were below cost of production. The reason for this anomalous behavior was that most of the large manufacturers were heavily bonded and had to meet debt obligations. *Lumber World Review*, July 25, 1912, 25. Robert Fullerton attributed oversupply to year-round operations in the South and on the Pacific Coast. In the old days snow and ice had kept mills in the North idle during the winter months, and supply had tended to stay in line with demand.

[7]*Northwestern Lumberman*, April 10, 1886, 2, Aug. 18, 1888, 1; Larson, *op. cit.* 261-262.

[8]*Northwestern Lumberman*, Aug. 18, 1888, 1, Sept. 13, 1890, 1; 26 *U. S. Statutes at Large* (1890) 582-583; Larson, *op. cit.* 262.

[9]*Southern Lumberman*, Feb. 15, 1892, 2-3. An editorial writer in the *Southern Lumberman* said that the proposal to put lumber on the free list threatened the southern lumber interest because as soon as duties were removed, American lumbermen in northern states adjacent to the Canadian border would erect mills across the border and thus increase the supply of Canadian lumber in Chicago and other nearby cities where a large part of the yellow pine production was consumed. In 1893 Canadian manufacturers agitated for a two-dollar export duty on logs as a means to compel the United States to repeal the one-dollar import on rough lumber. According to Canadians the one-dollar duty restricted lumber exportation to the United States and at the same time stimulated the export of logs to mills in the northern United States. Those interested in the speedy development of the yellow pine industry favored a high tariff on foreign-produced lumber and hoped Canada would put an export tax on logs. Such action would encourage millmen in the Lake states to establish mills in the yellow pine country and would also preserve the northern consumer market for southern made lumber. *Southern Lumberman*, May 1, 1893, 3; 28 *U. S. Statutes at Large* (1895) 545-546; Larson, *op. cit.* 262.

[10]30 *U. S. Statutes at Large* (1897) 167.

[11]Another witness took issue with Hines concerning the relative values of stumpage in the southern United States and Canada. He asserted that Canadian stumpage was disposed of at prices in excess of those in the United States. Since 1888 the Canadian government had reserved fifty cents per thousand on all timber cut on Crown lands, and timber was sold to highest bidder. In Ontario stumpage values were $6 to $7 per thousand board feet. Most of the timber in British Columbia was held under title called special license. Such licenses gave lumbermen the right to cut timber on Crown lands upon payment of fifty cents per thousand board feet stumpage and annual dues of $140 per square mile on lands west of the Cascade Mountains and $115 on lands east of the mountains. Licenses were reviewable annually for no longer than twenty years.

[12]*Tariff Hearings*, 2927-2935, 3069; *Southern Lumberman*, Feb. 20, 1909, 35. The better grades of southern lumber met no competition from Canadian woods, but 25 per cent or more of southern production, derived from the upper part of the tree and from small young timber, fell in the lower grades. Lumbermen usually cut all trees above eight inches in diameter at the top, since they did not expect to relog the land. Indeed, because the skidders would knock down much of the timber left by the sawyers, many lumbermen took every tree that would make a board. Had lumbermen followed conservative logging practices, cutting only the trees from which good grades of lumber were obtainable, the amount of low grade lumber produced would have been relatively small.

[13]U. S. Bureau of Corporations, *op. cit.* 63-64; 36 *U. S. Statutes at Large* (1906) 33; *Southern Lumberman*, Aug. 7, 1909, 28. In the opinion of F. W. Taussig free lumber would have increased slightly importation from Canada and checked to a limited extent the depletion of American forests. A few southern representatives voted for the retention of the duties on lumber, and their votes turned the scale in its favor. Taussig believed that because of the geographical limitation on competition and the different quality of southern lumber, the duty on Canadian lumber was of no real consequence to southern producers. F. W. Taussig, *The Tariff History of the United States* (New York 1923) 283.

Chapter XV

[1]*Twelfth Biennial Report of the Railroad Commission of the State of Mississippi* 1909 (Nashville 1910) 186-192.

[2]A. W. Nelson, Statement of the Mississippi Charter of the Society of American Foresters, 1954. (Manuscript prepared by A. W. Nelson of the Flintkote Lumber Company, Meridian, Miss.; copy in possession of the author.)

[3]Materials upon which narrative is based are derived chiefly from *Reports of the Railroad Commission of Mississippi, Interstate Commerce Commission Reports,* and the lumber trade journals.

[4]James Hutton Lemly, *The Gulf, Mobile and Ohio* (Homewood, Ill., 1953) 289-297.

⁵23 *Interstate Commerce Commission Reports* (1912) 290-292, 644; *Lumber Trade Journal,* Jan. 1, 1911, 23. Ordinarily lumbermen did not incorporate the spur logging roads because to do so would give other millmen the use of them and in consequence bring competitive bidding for the timberland tapped by the logging road. The fact was that a lumberman who owned unincorporated logging lines was able to control timber prices in the localities where he possesssed a transportation monopoly.

⁶*Twelfth Biennial Report of the Railroad Commission of the State of Mississippi* 1909, 177; 23 *Interstate Commerce Commission Reports* (1912), 285, 637-638, 644-645. Originally all rails used in a lumber operation were called logging roads. After the practice arose of extending to lumber companies a part of the freight division on products originating on logging roads, rails leading from the mill to logging camp or some other point became known as tap lines. Spurs radiating out from the tap line or main line continued to be called logging roads.

⁷32 *U. S. Statutes at Large,* Pt. 1 (1903) 847-848.

⁸Alfred H. Kelly and Winfred A. Harbison, *The American Constitution, Its Origin and Development* (New York 1948, 1955) 552.

⁹10 *Interstate Commerce Commission Reports* (1905) 511, 514-516; 206 *U. S. Supreme Court Reports* (1907) 1129.

¹⁰*Congressional Record,* 59th Cong., 1st Sess., Vol. 40, Pt. 7 (1906) 6889.

¹¹10 *Interstate Commerce Commission Reports* (1905) 198-212. It appeared to the Interstate Commerce Commission that the divisions received from trunk lines, though of great importance to the millmen, were insufficient to bear the entire cost of log transportation. It required about four carloads of logs to produce one of lumber. The expense of transporting logs from the forest to the mills was considerably greater than would have been the expense of carrying lumber from the same point. In some cases a division of freight rates was important to millmen. A thousand board feet of green lumber weighed on the average 4,500 pounds, and a similar amount of dressed lumber on the average 3,300 pounds. A division of three cents per hundred pounds amounted to a saving of about one dollar on each thousand board feet of lumber. One manufacturer stated that his division amounted to approximately $50,000 annually.

¹²10 *Interstate Commerce Commission Reports* (1905) 196. The Elkins Act declared deviation from the published rates and rebates to be unlawful. Deviation from published rates was defined as freight charges made by railroads on a commodity greater or lower than the official published rate. A rebate was giving by the carrier to the shipper a portion of the freight revenue on a commodity. 32 *U. S. Statutes at Large,* Pt. 1 (1903) 847-848.

¹³23 *Interstate Commerce Commission Reports* (1912) 644-647; *Lumber World Review,* June 10, 1912, 32-33.

¹⁴234 *U. S. Supreme Court Reports* (1914) 1193-1196.

Chapter XVI

¹Statements of Weston, Barber, and Baxter; Maxson, *op. cit.* 10.

[2]Tenth Census 1880, Schedule No. 3; Pascagoula *Democrat Star*, Feb. 22, 1889, July 9, 1887; statements of John Hickman and Barber.

[3]Pascagoula *Democrat Star*, March 8, 1889, March 15, 1889; *Southern Lumberman*, March 15, 1889, 6, April 1, 1889; Biloxi *Herald*, April 3, 1888.

[4]Pascagoula *Democrat Star*, May 18, 1900, May 25, 1900, July 20, 1900, Aug. 16, 1900; *American Lumberman*, May 19, 1900, 25, July 14, 1900, 27.

[5]Pascagoula *Democrat Star*, Feb. 6, 1893.

[6]Papers of Governor John M. Stone, Vol. II, in Mississippi State Department of Archives and History, Jackson, Miss.

[7]Lowe, *op. cit.* 133; Thirteenth Census 1910, *Manufactures*, 504. From this point on the source materials used were: *American Lumberman; Lumber Trade Journal; Southern Lumberman; St. Louis Lumberman; L. N. Dantzler Lumber Company Records; Mississippi Laws* (1904) 199-202, *Mississippi Laws* (1912) 165, 173-174; *Tariff Hearings*, 3042; U. S. Bureau of Labor Statistics, *Wages and Hours of Labor in the Lumber, Millwork, and Furniture Industries, 1890 to 1912*, Bull. No. 129, 33; statements of Barber, John Hickman, Hinton, and Simpson.

[8]*Southern Lumberman*, Oct. 25, 1913, 24; Dec. 20, 1913, 82.

CHAPTER XVII

[1]*The Constitution of the State of Mississippi 1890 and Amendments Subsequently Adopted*, compiled by J. P. Coleman and Heber Ladner (Jackson 1954) 70; *Mississippi Laws* (1892) 272; *Mississippi House Journal* (1898) 13, (1900) 23.

[2]Enochs, *op. cit.* 6-7.

[3]*Mississippi Senate Journal* (1906) 1091-1092.

[4]*Mississippi Laws* (1906) 283.

[5]*Southern Lumberman*, July 2, 1910, 27.

[6]The lumber journals previously cited gave extensive treatment of the Hines case; 106 *Mississippi Reports* (1913-1914) 804.

[7]*American Lumberman*, Sept. 1906, 1, Feb. 6, 1910, 23; Stone County Archives, Tax Rolls 1912-1930; L. N. Dantzler Lumber Company Records, Tax Records.

[8]*Mississippi Laws* (March 18, 1908) 56; Stone County Archives, Tax Rolls 1912-1930. In parts of Stone County taxes on standing timber moved steadily upward in the years 1916-1924. The tax history of one section of virgin timberland, of which one half (320 acres) was located in school district A and one half in district B, is contained in the figures below.

Date	Acreage	Assessed no. of board ft.	Valuation	Tax rate (mills)	Tax per 1000 bd. ft.	Tax per acre of land
SCHOOL DISTRICT A						
1916	320	1,480,000	$ 3,310	28	$ .063	$ .32
1919	320	2,600,000	13,000	37	.19	1.50
1924	320	2,600,000	14,660	59	.33	2.70

DISTRICT B

1916	320	1,750,000	$ 3,400	27	$ .05	$ .29
1919	320	2,945,000	14,200	28.75	.14	1.28
1924	320	2,945,000	16,080	39	.21	1.95

[9]*134 Mississippi Reports* (1916) 825; *Southern Lumberman*, Dec. 5, 1908, 24; *Lumber Trade Journal*, Jan. 1, 1909, 12-13; *Tariff Hearings*, 2920-2930.

[10]*St. Louis Lumberman*, Feb. 15, 1908, 20; *Southern Lumberman*, Feb. 1, 1908, 59. The U. S. Forest Service in collaboration with the State Geological Survey issued in 1908 a report which showed that only a very small percentage of South Mississippi timberlands was suited for agriculture. Such was the opposition of lumbermen, commercial clubs, and Mississippi representatives in Congress to the original finding that the report was subsequently revised to show a much higher percentage of pinelands adaptable to profitable agriculture. *Lumber Review*, April 15, 1908, 28; *St. Louis Lumberman*, April 15, 1909, 39. The Annual Report of the Forestry Commission in 1933, however, gave a fairly accurate picture of agriculture in the pine country. In none of the twenty-two longleaf counties was the percentage of the total area in cultivation as high as 30 per cent, and in only three was it above 20 per cent. Jefferson Davis County, 27 per cent in farmland, ranked first. Stone, Harrison, Hancock, and Jackson counties were lowest with only 2 per cent of their lands in farms. *Fifth Annual Report of the Mississippi State Forestry Commission 1933*, 50.

[11]*Tariff Hearings*, 2994.

[12]*Mississippi Laws* (March 8, 1912) 169-170; (March 27, 1918) 160.

[13]Statements of Weston and John Hickman; observations of the author in Forest, George, Jackson, Stone, and Perry counties.

[14]Statements of Barber, Batson, Weston, and Tatum. The Tatum family today own one of the largest tracts of timbered land in south Mississippi. Mayers Dantzler credits the survival of the Dantzler Company to Posey Howell's work in reforesting company lands. Howell started before 1920 the practice of leaving one seed tree on each acre of land, and he also limited the sizes of trees cut in lumbering operations. According to Barber and Dantzler, Howell met great difficulty in his efforts to convince both his employers and local residents of the need to restore the forests on the cutover lands.

[15]*Mississippi Laws* (March 26, 1926) 248-252.

[16]Charles Shotts, History of the Mississippi Forestry Commission (25 p. manuscript, 1954, in possession of present author).

[17]Mississippi Forest Industries Committee, *Mississippi Forestry Facts*, 7, 10; Southern Pine Association, Southern Pine and Total Lumber Production, 1869-1950 (one-page tabular statement, prepared in 1952, in possession of present author).

INDEX

Printed in the United States
140239LV00004BA/2/P

9 781604 732870